Sustainability and Innovation

AF167449

Editor-in-Chiefs

Jens Horbach

Series editors

Valentina de Marchi
René Kemp
Marco Lehmann-Waffenschmidt
Arthur P. J. Mol
Rainer Walz

Technological, institutional and social innovation promotes economic development and international competitiveness, and can do much to reduce environmental burdens. Thus, innovation is an essential factor in the realisation of the principles of sustainable development.

The series 'Sustainability and Innovation' aims to positively contribute to a better understanding of the connection between innovation and sustainability. It is interdisciplinary in coverage, concentrating primarily on the links between environmental and innovation research. Individual titles in the series focus on topical issues and cover both theoretical and empirical aspects: concepts of major practical relevance are also discussed. The series is addressed to researchers and policy makers, as well as to specialists and leaders in business and industry.

More information about this series at http://www.springer.com/series/6891

Pascal da Costa • Danielle Attias

Editors

Towards a Sustainable Economy

Paradoxes and Trends in Energy and Transportation

 Springer

Editors
Pascal da Costa
Ecole CentraleSupélec
Gif-sur-Yvette, France

Danielle Attias
Ecole CentraleSupélec
Gif-sur-Yvette, France

ISSN 1860-1030   ISSN 2197-926X   (electronic)
Sustainability and Innovation
ISBN 978-3-030-07717-4   ISBN 978-3-319-79060-2   (eBook)
https://doi.org/10.1007/978-3-319-79060-2

Printed on acid-free paper

This Springer imprint is published by the registered company Springer International Publishing AG part of Springer Nature.
The registered company address is: Gewerbestrasse 11, 6330 Cham, Switzerland

Foreword: Fresh Challenges to Overcome

The early twenty-first century has seen four fundamental and transformative changes in the economic organization of the world:

- Globalized trade has resulted in the production of goods and services now being shared internationally, with large countries like China and India opening themselves up to the market economy, following in the footsteps of the Asian *Tigers*.
- The development of digital technologies has transformed the planet into a *global village* connecting people, for better or worse.
- Financialized economies have resulted in all human labor now being assessed in terms of market value, thus eradicating incalculable labor.
- The impact of climate change, environmental degradation, and the depletion of natural resources could lead humankind down the path of potential destruction.

One of the consequences of these four factors is a shift in relative weights of global regions. The so-called industrialized countries' share of GDP is shrinking, while the least developed countries are now registering the highest growth rates.

Access to the resources necessary for economic activity is a further concern, as are the attendant greenhouse gas emissions, where the situation is constantly deteriorating. Tensions observed in 2007–2008 have somewhat eased as a result of the global financial crisis, slowing down Western economies and drawing supplier countries into this downward spiral. Without drastic changes, however, these tensions are bound to rapidly return.

This new situation will undoubtedly involve more violent reactions from populations aspiring to the greater well-being brought about by economic and social development, including in developed countries. Climate will be a major component of this well-being because of its dramatic impact on the lives of millions of people living in areas threatened by rising water levels, desertification, and hurricanes. The inevitable result is decisions imposed by force.

In view of these risks, what are the options? We need to recognize that each and every individual has equal dignity and a right to full development in all spheres: material, cultural, and spiritual. The world of the future would therefore give everyone access to the conditions for this essential human dignity. Once individuals are free, society must be organized to allow them to exercise their creative freedom and give all people the opportunity to develop their talents.

This freedom and dignity requires that individuals everywhere have access to work carried out in decent conditions and are paid sufficiently to support families. Demographics although reveal that increasing numbers of people are in the situation of having to work, due to population growth, and enter retirement at a later age, as a result of longer life expectancy. The only way to provide work for all is to maintain continued economic growth, and more particularly to promote the development of small and medium-sized enterprises, which grow more easily than very large or very small companies. However, this growth will generate polluting environmental disturbances and engender the further depletion of natural resources.

In particular, energy demand will continue to increase globally, and the sources of this energy must imperatively be decarbonized. In the field of transport, electric mobility will become increasingly the standard, requiring the use of new technologies. Paradoxically these technologies will use mineral resources whose stocks are poorly documented and whose extraction conditions will inevitably raise social and environmental issues.

Human beings naturally interact with each other. This sociable characteristic involves working for the common good and the collective realization of a fairer and more fraternal world. Humans are naturally concerned with the future of the planet and humankind and with finding the means to make the development of the poorest compatible with the preservation of everything around us, i.e., the environment in its widest sense, which includes culture.

The work of Pascal da Costa and Danielle Attias is mainly devoted to finding solutions for these dilemmas. These issues, be they ecological, social, or financial, will inevitably become the concern of companies, consumers, and public authorities. The latter will be able to use various levers ranging from infrastructure investments to binding measures, along with a host of incentives and different forms of economic regulation.

The desire to make works like these available to the greatest number, ideally everyone, motivated me to create the endowment fund CapitalDon in 2011. The fund's purpose is to financially support teams of researchers to dig deeper into all themes related to this issue. CapitalDon therefore naturally supported the work leading to the publication of this book.

I am convinced that it is possible to design an economic system based on giving meaning to companies, and ensuring that meaning is useful for society, and to call on everyone to actively participate in the challenges of progress, to ensure the sustainability of the global economy and the durability of enterprise. Business leaders, their employees, and all company stakeholders are in the same boat, as we are all in the same planetary vessel. Each person is, on his or her level, responsible for the success of civil society as a whole. And the only way to ensure these results is to make sure that it is humane, in accordance with freedom and with respect for dignity.

Founder of CapitalDon   Pierre Deschamps
CapitalDon, Paris, France

Acknowledgments

This book is the result of research carried out over several years within the Economics and Management "Sustainable Economy" Group at CentraleSupélec. We would like to extend our gratitude to Mr. Pierre Deschamps, patron of the Chair of Sustainable Growth Capitaldon, for his support and his confidence. Our acknowledgments also go to the esteemed researchers at the Industrial Engineering Laboratory at CentraleSupélec for the numerous fruitful exchanges we had and finally to Ms. Divya Madhavan for her invaluable assistance in the writing process.

Contents

About the Editors

Danielle Attias is Professor at Ecole CentraleSupélec (France), Chairman of Research Chair of Armand Peugeot, Hybrid Technologies and Economy of Electro-mobility, Member of Sustainable Economy Team Research in the Laboratory of Industrial Engineering of CentraleSupélec, and Expert in the European Research Center for Road Traffic Safety.

Her research topics are the economics of innovation and electro-mobility.

She has published a book, *The Automobile Revolution: Towards a new Electro-mobility Paradigm*, with Springer, 2017, and seven articles in refereed journals. She is a current and former member of 11 scientific committees and also member of the French Parliamentary Commission on Mobility. She is currently supervising three theses within the framework of the Research Chair.

Pascal da Costa is Associate Professor at Ecole CentraleSupélec (France), Chairman of the Research Chair *Croissance Durable Capitaldon*, and Head of the Sustainable Economy Team Research in the Laboratory of Industrial Engineering of CentraleSupélec.

His research topics are the economics of sustainable growth and energy.

He has published in 11 refereed journals, three articles in revision, 20 chapters in books, and three editing roles in books. He is current and former member of nine scientific committees, participating in five ANR projects (French Research Agency) and four EU research programs, supervising eight theses, six postdoctoral researchers, and nine master's students.

Introduction: Reframing the Notion of Sustainable Economy Through Perspectives on Limits, Tensions and Paradoxes Within the System

Pascal da Costa and Danielle Attias

Abstract The ambition of the book is first to clarify the concepts of sustainability or sustainable economy, to study their most diverse fields of application, and then to display their limits and highlight the very many tensions that exist between them.

The core themes of the book are the paradoxes and trends related to the future sustainability of economic systems, sectors and firms, in the context of limited natural resources. The ebb and flow between theory and practice proves essential towards building a better understanding of current contradictions of the sustainable economy.

Keywords Sustainable economy · Electricity grid and generation · Automobile technology and electric batteries · Life cycle assessment · Urban transport policies · Smart cities · Transformation

Sustainable economy is represented in the interactions between the present and future of economic, social and environmental issues—and is often equated with sustainable development. These are indeed essential notions for all global economies and have long been the subject of numerous reports, studies and analyses (WCED 1987; Meadows et al. 1972, 1992; Daly and Townsend 1993; Pearce and Atkinson 1995; Spash 1999; Tahvonen 2000; Jackson 2009, 2014, etc.). Is there then anything that remains to be said on this, albeit crucial, concern? Is the characterisation of the interdependencies between natural and human systems within a systemic view even possible? *"Is the environment changing the nature of the economy?"* as argued by Laurent and Le Cacheux (2012).

Economics and ecology have the shared etymology from the Greek *oikos*—and are both concerned with the interactions between man and his environment: do economics and the management of sustainability account for the complexity of environmental issues and their attendant policy choices?

P. da Costa (✉) · D. Attias
Laboratoire Genie Industriel, CentraleSupelec, Université Paris-Saclay, Gif-sur-Yvette, France
e-mail: pascal.da-costa@centralesupelec.fr

© Springer International Publishing AG, part of Springer Nature 2018
P. da Costa, D. Attias (eds.), *Towards a Sustainable Economy*,
Sustainability and Innovation, https://doi.org/10.1007/978-3-319-79060-2_1

The ambition of this book is twofold: first to clarify these concepts and study their most diverse fields of application, and then to display their limits and highlight the very many tensions that exist between them. The core themes of the book are the paradoxes and trends related to the future sustainability of economic systems, sectors and firms, in the context of limited natural resources. The ebb and flow between theory and practice is essential towards building a better understanding of current contradictions of the sustainable economy. The observations made by Theys (2002) allow us to position the terms of our debate: *"Undoubtedly, the concept of 'sustainable development' is distinguished by a remarkable capacity to pose and, above all, link central questions facing our societies: the question of the goals of growth—and the possible compromise between the divergent interests of economy, society and ecology; that of "time" and of short-term and long-term competition, present and future generations; That of "spatial identities"—and the problematic articulation between the logics of globalization and automation of local territories"*. A global approach, accompanied by multidisciplinary reflection on this subject is however not very present in existing literature. From our point of view, it is necessary to highlight the numerous questions raised by this economic model and its objective limits.

The most distinctive feature of this book is its interdisciplinary academic research covering a wealth of issues, topics and methods, towards a more illustrative and narrative description of what lies down the long road to sustainability. Our book deals with government regulation, management strategy and company responsibility. These aspects are analysed mainly in the domains of energy and transport- drawing on complementary methods from the engineering and management sciences as well as economics, within the overall framework of a systems sciences approach.

The chapters of the book foregrounds issues on prospective advances in various fields (electricity grid and generation, automobile technology and electric batteries, etc.). Our aim is to shed light on how these technological systems might contribute to the transformation of the economic system. These systems are broadly construed as eco-innovation, and also known as the famous 'transition' in the field of energy, towards sustainable paths of growth, towards the 'green economy'.

The first part of the book seeks to examine government CO_2 targets. These targets are negotiated under the United Nations Conventions on climate change. The issue of regulation and the future deployment of renewable energies or electric vehicles are analysed within this framework, yielding a comparison of energy policies in European regions with reflection on the state-market dilemma.

The electricity sector in Europe clearly raises this question, providing a powerful example of a sector facing issues of sustainability within global climate change, with its possible utopias and probable paradoxes. In the first chapter (i.e. Chap. 2 of the book), we demonstrate that the drivers of investment decisions related within the domain of electricity capacity in Europe have undergone considerable transformation in the post-war period, from 1945 to the present. In this context, the differences between rational behaviors, as advanced in the theory, and actual investor behavior and government action are highlighted. As European Union (EU) climate policy has intervened in the electricity market -due to the influence on prices at multiple levels, with consequences for producers and consumers. The assertions of liberalization policy and climate policy must be constructed. These were initially separate packages in EU policy and were thus not harmonized until recently. In this vein, we examine the movement of recent re-centralization of European energy policy, in the form of new regulation related to climate and renewables.

To complement the qualitative approaches, the book includes original quantitative studies, such as a systems analysis of technological roadmaps (Chap. 3). Many countries participating in COP21 (Paris, December 2015) have since developed technology roadmaps—which include the transformation of the electricity sector—towards achieving their national targets for reducing CO_2 emissions. In a global warming context, these roadmaps take into account power, transport, housing and the industrial sectors, as well as major technological advances such as carbon capture and storage, energy efficiency and electric vehicles. Using an original econometric model, we show that the massive deployment of electric vehicles remains indispensable for reaching government targets in France, China and the US. The ability of these economies to reach CO_2 reduction targets will necessarily entail drastic changes in the transportation sector. A few of the technology roadmaps simulated by our model demonstrate the following: in France, 80% of vehicles must be exchanged for electric vehicles and energy efficiency in residential housing and industries must be improved by 80% by 2050. In the United States, at least 80% of cars must become electric, residential and industrial energy efficiency must be improved by at least 60%. Note that in the power generation sector, the use of coal in 2050 can be at most half of 2010 levels. In China, if electric vehicles can replace 60% of cars, then energy efficiency would have to be improved by 60% and coal use reduced by 60% by 2050.

Issues of energy resources are at the heart of this book. We study the paradoxical use of non-renewable mineral resources in renewable energies and the electrification of the economy. The issue of future management of our natural resources is upon us, as are energy choices.

Chapter 4 discusses the electric vehicle (EV) industry, which remains in the introductory stages of the product life cycle, where the dominant design still is unclear. Electric vehicle companies, both incumbents from the car industry as well as newcomers, have for long made efforts to promote electric vehicles in the niche market by offering innovative products and business models. While most carmakers still take a business-as-usual approach to developing their EV production and products, Tesla Motors stands out for its disruptive innovation. This chapter provides a literature review on the business model approach and a classification of innovations in the electric vehicle ecosystem. The lessons from this chapter for more traditional original equipment manufacturers (OEMs) in designing their business models for electric vehicles would merit their attention if Tesla's disruptive choices succeed in challenging the dominant design.

Chapter 5 focuses on the scenario of Europe: we look at the availability of constituent materials and the impact of recycling on electric vehicles in Europe, towards assessing the potential for critical shortages resulting from the scaling up of electro mobility. Lithium-ion battery technology represents a key component of vehicle electrification. End-of-life recovery of these batteries is an important factor in lifting the barriers to increased electro mobility. These include battery cost and environmental impact, as well mandatory recycling rates of more than 50% of battery weight (European Union regulation) and, finally, the availability of constituent elements such as lithium and cobalt. Our analysis shows that recycling significantly reduces the consumption of materials used in lithium-ion batteries.

The aim of Chap. 6 is to present the Life Cycle Assessment (LCA) method. LCA is an approach to assessing environmental performances of products throughout their life cycle and to provide an integrated framework to identify the best fit for a specific context, considering environmental, economic and commercial aspects. LCA-based decision-making usually focuses on environmental impact, excluding other consideration such as customer expectations and economic aspects. The framework is applied on three burners for forge furnaces. Results show that client profiles and operating contexts (namely client expectations, location, resource availability and costs) have a strong effect on technological choices.

The book then focuses on the implementation of government urban transport policies, either through incentives (such as inter-modality and taxation) or coercion (such as speed limits and urban tolls). The CO_2 reduction target set by the European Commission is one such example. In the transport ecosystem, the incentives and binding rules imposed by public authorities are determinants, since they affect the choices of all economic actors, facing us with a new paradigm. By 2050, 70% of the world's population will live in or around cities, cities already generating 70% of energy-related greenhouse gas emissions. The future of urbanisation will be smart, with optimised land use and a transport system that is more efficient and environmentally friendly.

In a smart city, urban and transport planning should be co-conducted harmoniously in order to create a new transit-supporting city. After defining our vision of smart mobility, we will present and analyse the links between the transport system, disruptive innovation, and the role of public policies in change management in Chap. 7. We focus on how the co-conception of smart mobility, defined as disruptive eco-innovation, is organised in a local territory. The development and diffusion of innovation within the mobility ecosystem significantly disrupts usages and modifies market boundaries. Implementation conditions to achieve a widespread adoption of smart mobility are discussed and the role and decision-making methods of territorial actors are considered. This part concludes with an examination of the role of governments and local actors in the transformation of the automotive industry into an ecosystem of electro mobility and how this transformation is taking place by placing the government's key players in transport challenges, in a new structuring role.

Chapter 8 focuses on the role of governments in the transformation of the car industry into an electro-mobility ecosystem and how this transformation is taking place by placing key government players in transport challenges, endowed with a new structuring role. Partnerships between private and public actors are necessary, although complex. Combining public and private offers for this new mobility creates opportunities, but also constraints. The revolution in urban mobility aims to be intelligent and user-centred. This concept of mobility-as-a-service based on an offer of mixed mobility will afford the city-dweller safer and more sustainable mobility. We reveal the paradoxes of public policies in the transition from the old transport paradigm based on the proliferation of cars-associated with multiple nuisances-to a fresh paradigm embracing new needs and providing transport solutions in a sustainable environment.

We conclude this book with a chapter that opens up new perspectives through an anthropological approach to transformation, inspired by Thompson (1917). Our research also consists in attempting to understand and capture forms. Forms refer to the idea of transforming economic systems, while evaluating their similarities and differences as well as their continuities or invariants through time. Organizations and firms feel the need to define their structures, the shape of their relationships and interconnections. This raises the interdisciplinary research question: Can we think about the transformation of functions that do not have any intrinsic shape? By no intrinsic shape we mean not only new or poorly defined forms, but also forms that are not yet established, nor shared by stakeholders and poorly understood. This is typically the case in sectors that have been dramatically changed by sustainability, new competition and regulation, to the point that they are no longer defined by their past.

This book contains a wholly unique multidisciplinary approach that meshes qualitative and quantitative studies, some scenario-based, others based on econometrics and data. The chapters deal with real cases of individual organizations (Tesla, Navya, etc.) production chains (batteries, etc.) and macroeconomic studies of national economies (electricity, technological roadmaps, etc.). This range of approaches allows us to put forward an inventory of sustainability in ways that are entirely distinctive from the existing literature.

References

Daly, H. E., & Townsend, K. N. (1993). *Valuing the earth: Economics, ecology, ethics*. Cambridge MA: MIT Press.

Jackson, T. (2009). *Prosperity without growth*. London: Sustainable Development Commission.

Jackson, T. (2014). The dilemma of growth: Prosperity v economic expansion. In *The guardian 'rethinking prosperity'*. Guildford: University of Surrey.

Laurent, E., & Le Cacheux, J. (2012). *Économie de l'Environnement et Économie Écologique*. Paris: Armand Colin.

Meadows, D. H., Meadows, D. L., Randers, J., & Behrens, W. (1972). *The limits to growth: A report for the club of Rome's project on the predicament of mankind*. New York: Universe Books.

Meadows, D. H , Meadows, D. L., & Randers, J. (1992). *Beyond the limits: Global col- lapse or a sustainable future*. London: Earthscan.

Pearce, D., & Atkinson, G. (1995). Measuring sustainable development. In D. W. Bromley (Ed.), *The handbook of environmental economics* (pp. 166–181). Oxford: Blackwell.

Spash, C. L. (1999). The development of environmental thinking in economics. *Environmental Values, 8*, 413–435.

Tahvonen, O. (2000, June). *Economic sustainability and scarcity of natural resources: A brief historical review*. Washington, DC: Resources for the Future.

Theys, J. (2002). L'approche territoriale du développement durable, condition d'une prise en compte de sa dimension sociale. In *Développement durable et territoires*, Dossier 1: Approches territoriales du Développement Durable.

Thompson, D'A. W. (1917). *On growth and form*. New York: Cambridge University Press.

WCED. (1987). *Our common future*. New York: Oxford University Press.

Part I
From Energy Market Regulations to CO$_2$ Targets

The Paradoxes of the European Energy Market Regulation: A Historical and Structural Analysis of the Electricity Mix

Bianka Shoai-Tehrani and Pascal da Costa

Abstract The aim of the Chap. 2 is to understand how the drivers of investment decisions in electricity production have evolved over time-from 1945 to the present day, in the specific context of Europe facing wars and conflicts, scientific and technological progress, all within environments undergoing strong political and academic developments.

We study investment in power production decisions by comparing the history of European electricity markets with successively dominant economic theories in this field. Therefore, we highlight differences between rational behaviours, such as those described by theory, and actual behaviours of investors and governments. Liberalization is clearly on the agenda given its 25-year history in terms of European Union markets, as well as forming part of a rationalization that is prescribed by new economic theories. It remains considerably heterogeneous, which complicates the creation of a large single market for electric power within the Union.

We see also new constraints on energy policy in Europe, which takes the form of new regulation, mainly relating to climate and renewables. As liberalization and climate policy were initially separate packages in EU legislation, their combined effects pose a critical 'missing money problem' to major utilities, thus making for this re-regulation, that is nonetheless different from the centralized control experienced by all European electricity markets until the mid-1980s.

Parts of this paper were published in USAEE/IAEE conference proceedings in Open Access: http://www.usaee.org/usaee2016/submissions/OnlineProceedings/2060-USAEE%20Tulsa%20Full%20Paper%20Shoai.pdf (Shoai Tehrani, B., Da Costa P., Akimoto K., Nakagami Y. (2016). Are Deregulated Electricity Market and Climate Policy compatible? Lessons from overseas, from Europe to Japan, proceedings of the USAEE).

D. Shoai-Tehrani (✉)
Systems Analysis Group, Research Institute of Innovative Technology for the Earth (RITE), Kizugawa, Japan
e-mail: shoai@rite.or.jp

P. da Costa
Laboratoire Genie Industriel, CentraleSupelec, Université Paris-Saclay, Gif-sur-Yvette, France
e-mail: pascal.da-costa@centralesupelec.fr

Keywords European electricity market · Electricity investments · European energy market liberalisation · Climatic issues · Renewables

1 Introduction

This chapter seeks to understand investment decision drivers in electric power capacities and their evolution over time; from 1945 to present day, under different regulatory schemes in Europe. We will address this issue from theoretical and historical perspectives, within the wider context of a transformative academic, political and scientific landscape nestled in the post-war period. This period has witnessed considerable evolution in drivers for investment decisions.

Today, electricity investment is subject to transformation resulting from ongoing European market liberalization and, more recently, breakthroughs on climate change, the latter imposing reductions on greenhouse gas (GHG) emissions, in particular through planned integration of renewables in the generation mix. As EU climate policy interfered with the electricity market through the influence on prices at several levels, with consequences for both the producers and consumers, the dialogue between liberalization policy and climate policy (that were initially separate packages in EU policy and thus unharmonized) must be reconsidered and built.

In this regard, the main goal of such dialogue would be the evolution of the generation mix towards low carbon electricity. The liberalisation, now clearly under question, initiated a process of rationalization prescribed by 25-year-old economic theories of the 1980s and 1990s. This has indeed yielded heterogeneous results regarding market structure, prices or power quality (i.e. continuity in production and sufficient generation). There has also been a *re-centralization* with energy policy more recently, in the form of new regulation regarding climate and renewables, and of programmed investments in grid interconnections for EU member countries. As liberalization and climate policy were initially separate packages in EU policy, their combined effects lead major towards a critical 'missing money problem', mandating re-regulation. This *re-regulation* is different from the familiar centralized drivers in Europe until the 1980s, as it does not question the liberalization process per se. It aims at allowing heavy investment and for instance, providing more support to the electricity market through new taxation tools (carbon tax, research and development subsidies...) aimed at *internalizing external effects* and reconciling liberalization and climate policy in an integrated policy.

Two main historical periods tease out and structure the two first parts of the chapter: the 1945–1986 period, during which national generation mix takes shape in European countries, often in contradiction with one another in terms of economic optimization, privileging local resources, Ramsey-Boiteux rule, etc. (Sect. 2); The 1986–2016 (current) period, marked by important transformations: the objective of liberalizing the electricity sector ending up in different degrees of competition in EU countries; new climate stakes and recent development of renewables (Sect. 3). The third part of the chapter identifies current challenges and future trends facing the electricity sector for new investments, as climate change issues gain more and more importance (Sect. 4).

2   1945–1986, from European Reconstruction to Oil Shocks: A Crucial Period for the Constitution of Current Power Generation Mix

2.1   Nationalization or Integrated Model?

The post-war period, saw reconstruction as a primary goal for all European countries. For the electricity sector, the priority was to go back to previous levels of generation as rapidly as possible. To do so, governments took measures that end them up with increased control on the electricity sector.

In France, the nationalization of power company was voted in 1946, which led to the creation of *Electricité de France* (EDF) (Beltran and Bungener 1987). In the United Kingdom, nationalization was, likewise, decided on according to the *Electricity Act* voted in 1947. The *British Energy Authority* was created in 1948 and became the *Central Electricity Generating Board* (CEGB) in 1957 (Grand and Veyrenc 2011). Italy also chose to nationalize the electricity sector in the Constitution in 1946, but national operator *Enel* was created only in 1962 (Grand and Veyrenc 2011) due to industrial reluctance in the sector: nationalization indeed means that *Enel* would have had to absorb the 1270 historical power operators. The process goes again and is then completed in 1995 (*Engel (spelling?) website*). In these three countries, the electricity sector has thus become a state monopoly. Governments have direct control over pricing and technology choices.

The situation in Germany and Spain was different: they do not create state monopolies nor centralized planning (Grand and Veyrenc 2011; Ibeas Cubillo 2011). The German electricity industry corresponds to an integrated model: including in its structure local and regional companies, due to the particular configuration of the German federal state itself—being divided in powerful *Länder*. Yet the sector is highly integrated on both vertical and horizontal scales through numerous exclusivity contracts not only between power generators and grids, generators and distributors, but also from generator to generator. In the end the electricity sector in Germany was not submitted to competition and the 1935 Energy Act maintained direct control over prices. Technological choices were however adopted at a federal level. In Spain, electricity sector integration happened through the coordination of private companies among themselves (Ibeas Cubillo 2011). In 1944, 18 electricity companies created the *Asociación Española de l'Industria Electrica (UNESA),* in order to promote a real national electricity grid by developing more interconnections to ensure better supply (Asociacion espanola de la industria Electrica 2013). Similar to Germany, the Spanish government controlled prices indirectly through the *Unified limited rates system* established in 1951 that set maximum prices and regular tariff harmonization in different areas of the country.

European states thus took control of the power industry either through a monopoly referred to as a "natural monopoly" by economic theory, or through an integrated model where potential entrants and prices were influenced by the state.

2.2 Cost-Benefit Analysis: The Dominant Economic Theory of the Period

In the aftermath of World War II, Cost-Benefit Analysis was the dominant theory for electric power investment all over Europe. It justified and supported the settling of monopolies and integrated markets. This theory was issued by works of *marginalist* economists and stems from the Welfare Economics of the 1930s and 1940s by Allais, Hicks, Pigou, Samuelson (Pigou 1924; Hicks 1939; Allais 1943; Samuelson 1943). In the 1950s, Cost-Benefit Analysis was initiated in France and other European countries by Boiteux and Massé (Massé 1953; Boiteux 1956). This analysis entails assessing explicitly the total expected costs and total expected benefits for one or more electricity investment projects, in order to determine the best or most profitable.[1]

Technically, electric power supply at the time relied on two technologies: hydroelectric plants and thermal plants. Debates on the profitability of both technologies lead to important conceptual breakthroughs, a complete cost assessment of technology in particular—which also included lifecycle analyses for the facilities, choice of corresponding discount rates, and the ability of supply to match peak consumption.

Power generation per se is capital-intensive due to necessary grid and plant investments. This is why Cost-Benefit Analysis needs to be applied with integrated markets or monopolies, the latter referred to as '*natural*' according to the *Ramsey-Boiteux rule*. This rule demonstrates that a company with initial fixed costs (such as in the electricity sector) undergoes losses if its price is equal to marginal cost (perfect competition); whereas in a natural monopoly, it can reach equilibrium thanks to second order pricing superior to marginal cost and inversely proportional to demand elasticity (Boiteux 1956).

---

[1]The first optimization model based on Cost-Benefit Analysis was developed in 1955. Massé said at the time: 'The electricity industry has found a purely objective tool in order to take investment decisions without personal bias (Beltran and Bungener 1987).

2.3 The Lack of Risk and Uncertainty Assessment in Cost-Benefit Analysis

In Massé's works for optimal electricity investment determination, the main risks at stake are discussed. It is yet clear that they are insufficiently integrated in the modelling or only in a rather limited capacity (Massé 1953).[2]

After the Suez crisis in 1956[3](Chick 2007) the first weaknesses of Cost-Benefit analysis were clearly identified; exogenous risks such as supply risk on imported oil as was the case in the Suez crisis and its cascading effects were not correctly anticipated in this theory (Massé 1953; Denant-Boèmont and Raux 1998).

Economic theories on risk are nevertheless developed at the same time. In the 1940s and 1950s, Neumann, Morgenstern, Friedman and Savage (Neumann and Morgenstern 1944; Friedman and Savage 1948) addressed the issue of the *decision maker's rationality when confronted to the risks at stake*.[4] This progress was however excluded from marginalist modelling for electricity investment.

2.4 The Initial Competition Between Oil and Coal

Oil and coal then became the two main resources for thermal power plants. European coal producers rapidly felt threatened and demanded protection against foreign oil imports. They argued that high risk resides in the political instability of the Middle East; jeopardising supply, transportation and prices altogether. Did domestic coal producers get any protection in the 1950s and 1960s from cheap foreign oil imports?

In France, EDF were under no obligation to use more coal than necessary. This was easy given that France had limited resources in coal compared to Germany and the UK. Indeed, in the 1950s and 1960s, coal production reached 100 million tons in

---

[2]To be more specific:

- The risks related to operational costs and especially fuel costs were assessed by using past data: no changes in future trends were considered;
- The risks related to investment costs were mainly due to construction risks associated with the land on which the plant was being built: it was considered as a mathematical expectation that was added to the investment cost as a security expense;
- The risks related to financing programmes (volatility of public decisions) were identified but not taken into account;
- The risks related to the expenses of financial compensation offered due to damages caused by plant construction gave us a first glimpse of the internalisation of externalities, but again no modelling was considered since it was too risky to be assessed.

[3]The conflict occurred between Egypt and an alliance formed by Israel, France and the United Kingdom, after the nationalization of the Suez Canal by Egypt, the canal being a strategic step for oil imports.

[4]Weisbrod, Arrow et Henry completed these theories in the 1960s and 1970s by addressing the issue of public decision in uncertain environment (Weisbrod 1964; Arrow 1965; Henry 1974).

Germany (133 million in 1957) and 200 million tons in the UK (197 million in 1960), whereas France's maximum production reached 59 million tons in 1958 and could never ensure self-sufficiency (National Coal Mining Museum 2013; Office statistique des Communautés européennes 2016). Moreover, marginalist economists (who did not take into account the supply risk) recommended reducing coal production in France and increasing oil imports.

Contrary to France, the United Kingdom and Germany, who had considerable resources in coal, took measures to protect domestic coal production. In the UK, the government created a tax on oil imports in 1962, banned Russian oil and American coal imports, and from 1963–1964, imposed quantified coal use targets to CEGB (Chick 2007). In Germany, such measures will occur later, after the oil shock, but are part of the same approach. (I'd like us to review the tenses for this section together, it jumps from historical present to perfect to past to future...maybe it's ok but my caffeine-deficient brain isn't processing it right now).

2.5 *From* Peak Oil *to Developing Alternative Technologies to Oil*

After the two oil shocks in 1973 and 1979, a transitory period began in Europe. In reaction to high oil prices, all countries took measures to reduce their dependency on the black gold, including France that had not made this choice from the beginning.

A predictable effect of *peak oil* is the return to coal for some electricity producers. This happened mainly in Germany, where the Kohlpfennig was established in 1974- a tax on electricity consumption, used to support domestic coal. In 1977 the *Jahrhundertvertrag* (literally 'the contract of the century') makes it compulsory for power generators to get part of their supply from domestic coal producers.

The search for substitutes then developed, being very different from one country to the other. For instance, the United Kingdom quickly started to explore the North Sea for new fossil resources, like gas, while France invested massively in civil nuclear energy.

Electro-nuclear programs thus developed in France and Europe: their successes or failures depending strongly on the resistance of national economies and companies to oil shocks, succeeding in strategic and industrial nuclear deployment and managing public acceptance (or even public support).

In France, the high cash flows for EDF allowed for limiting the impact of high oil prices on consumers (Francony 1979). EDF also managed to have low financial costs for the building of its nuclear fleet. For purely economic reasons, the choice was made to go with the American Pressurized Water Reactor (PWR) technology and buy the corresponding Westinghouse license in 1969 rather than French Graphite Gas Reactors developed by the French Commission for Atomic Energy (CEA). The French nuclear program (Plan Mesmer) is thus launched in 1974.

The United Kingdom adopted the opposite approach. The nationally developed Advanced Gas Reactor (AGR) was chosen for the nuclear program (Grand and Veyrenc 2011). However the program was then abandoned in the middle of the 1980s for want of competitiveness. An alternative program based on the Westinghouse PWR technology was then launched in 1982 but abandoned again after the building of only one reactor in 1988 (*Sizewell B*) due to cuts in public budget and drifting costs.

In Germany, the technologies chosen by the companies were Pressurized Water Reactors (PWR) and Boiling Water Reactors (BWR) developed locally by a *Siemens* subsidiary. Nuclear energy grew rapidly in Germany, although it was contested by the public from the start (which was not the case in France). Between 1980 and 1986, Italy built only four reactors and Spain five.

Besides, public acceptance of power generating technologies became more and more vital over the years. Local opposition for environmental protection first focused on coal, demanding that coal-fired plants are built outside cities. The phenomenon quickly reached civil nuclear, particularly in Germany. The opposition to the building of a nuclear plant in Wyhl in the 1970s, successfully led to abandoning the project in 1975 and becoming an example for all anti-nuclear movements (Mills and Williams 1986).

The rejection of coal-fired plants by one part of the European population was first addressed by the development of the first Combined Cycle Gas Turbine (CCGT) in the United Kingdom in 1991. This technology allowed for the building of facilities smaller than coal and nuclear plants, while ensuring high profitability. It gained favour by the end of the Cold War (in the late 1980s) as it meant direct access to cheap and abundant Russian gas—indeed Russia was in 1990 the world's foremost gas producer with 629 billion m^3 (Enerdata 2012). Electricity producers using CCGT thus achieved competitiveness on the market thanks to accepted and moderate investment along with cheap gas.

Such new entrants stimulated competition on the electricity markets until they were then integrated or monopolistic. However, the liberal mutation of Europe regarding electricity was more due to a combination of theoretical breakthroughs and political decisions.

3 1986–2016, from the Process of European Liberalization to Climate Change Mitigation Considerations: Towards a Mutation of Electricity Markets

3.1 Theoretical Questioning of Natural Monopolies

In the aftermath of World War II, Cost-Benefits Analysis had shaped electricity investment choices in numerous European countries, remaining the major approach until the 1980s, although already theoretically contested in the 1960s. These works first questioned the efficiency of monopolistic and integrated models, and identified their negative effects empirically. First, a tendency to over-capitalize was revealed—it was the Averch-Johnson effect (Averch and Johnson 1962).[5] The absence of competition also failed to encourage efficiency (Leibenstein 1966). Besides, the relationship between the regulator and the electricity sector could result in a protection of the interests of the monopoly rather than the interests of consumers (Buchanan 1975; Stiglitz 1976; Peltzman 1976).

This questioning goes further with Kahn, Baumol et Sharkey who address the issue of how to define a natural monopoly (Kahn and Eads 1971; Baumol 1977; Sharkey and Reid 1983). In a grid sector such as the electricity sector, they claim, a natural monopoly does not apply to the whole sector but only to activities related to grid management. Competition can thus be introduced in other activities of the sector, such as production and distribution, for consumer benefit. This argument was the one later raised by EU and was at the root of the liberalization process in grid industries.

In the 1990s, (Laffont and Tirole 1993) emphasized these results by showing that a monopolistic company had an asymmetrical relationship with the regulator. The company's interest was thus to take advantage of this situation regarding information on key points in order to increase their revenue.

3.2 From European Coal and the Steel Community to the Directive Relative to the Internal Electricity Market

With the political construction of the EU in the 1950s, several European Communities for trade and economy were created. In 1986, these communities led to the Single European Act and in 1996 to the creation of a single European electricity market—or rather to the creation of such an objective—thanks to the EU Directive on *common rules for the internal market in electricity*.

The United Kingdom was a model for this market reform. Chronologically speaking, it was the first European country to experience electricity market liberalization (Glachant 2000). The creation of an internal market in Europe had two goals. First, competition was expected to lower electricity prices for consumers. Second, a European market allowed for broadening the perimeter for resources in order to have better system optimization (Grand and Veyrenc 2011). In practice, the reform allowed member states to choose whatever measures they saw fit to meet the objectives. They could either open the market to new entrants, or stop controlling prices, or create an independent regulator for every activity open to competition (Newbery 1997; Perrot 2002).

---

[5]This effect measures the tendency of companies to engage in excessive capital accumulation in order to increase the volume of their profit.

Given the heterogeneity of institutions, markets and industries across European countries and given also the flexibility of European Commission Directives, the results ended up being highly heterogeneous.

The United Kingdom was historically the first country in Europe to deregulate its market from the mid-1980s, together with the United States of America on an international level. Today, we take stock of the initial results of this deregulation. The picture is a mixed one. Clearly, British deregulation has followed a specific process by starting from an integrated industry: sorting power plants according to technologies: *British Energy* took charge of nuclear power plants and *Centrica* of others. *British Energy* historically stayed into generation without engaging in downstream activities. The sales activity focused on a few big clients (companies) with the rest of the generation being supplied through independent marketers. The opening of the market to competition on different aspects of the value chain: generation and distribution. Grid networks have a mixed regime: they are regulated but were allowed to be owned by actors of the competitive market.

Liberalization certainly needed to evolve in order to take into account the necessity of ensuring investments in new capacities. Today, the United Kingdom seems to have to intervene directly on the market to ensure necessary electricity investments. The agreement between the British government and French company EDF for the building of two EPRs is one such clear example (Department of Energy and Climate Change and Prime Minister's Office 2013).

The liberalization of the Italian electricity market also rapidly delivered visible results. The Italian state was favourable to liberalization from the start, it immediately auctioned part of the assets of historical oligopolies in order to favour new entrants. However, the importance of the power company Enel on the Italian stage (28% of national generation) as well as the international stage showed that there was still a strong national champion; which was not the case in the UK.

The overall attitude of Germany towards liberalization seemed favourable to begin with, but the process quickly instated a reinforcement of state control on electricity operators, who were formerly used to auto-regulation.

The current structure of the German electricity market is dominated by four companies: *E.On* and *RWE* ensuring 60% of generation[6]; *Vattenfall* and *EnBW* 20%. The relative failure of electricity market liberalization in Germany can be partially attributed to the German state's will to protect the volume of national electricity generation. While Germany has abundant coal resources; coal is the cheapest fuel today, this higher price in Germany can be explained by strong penetration of renewables and high taxes on electricity prices.

Spain has adopted an attitude similar to Germany's: state control on prices, protection of historical operators (*Endesa* and *Iberdrola*) and strong support of renewables.

---

[6]E.on and RWE are historically *multi-utilities* and are very present on the gas market as well as the electricity market.

France is the country where liberalization was the least successful: there is one main operator regularly supported by French state policy in its application of European directives (the December 2010 NOME law, to which we will come back later). France thus avoided some of the mistakes of the integrated model. It did not protect coal in the 1960s when it was not competitive compared to oil, and chose in the 1970s the most profitable nuclear technology even though it was not the one developed nationally.

3.3 Current State of the Liberalized Market

3.3.1 Market Integration and Market Structure

Regarding market integration, the EC identifies positive trends in the current single market progress (European Commission 2014a), such as the fact that market coupling progresses, or that the unbundling of transmission system operators (TSOs) from vertically integrated energy groups can now be globally considered a success. 96 of about 100 transmission systems in Europe are now in compliance with EU legislation. Connections were achieved between Estonia and Finland, UK and Ireland; electricity interconnections between Sweden and Lithuania are now under construction. However, current interconnections are deemed insufficient, which is why the EC has now set a target of a 15% minimum of installed electricity capacity, arguing that interconnections will help market integration as well as emissions reductions and energy efficiency. Priorities are set on North-South electricity interconnections in Western Europe, such as improving the connection of Ireland, UK or the Iberian Peninsula with the continent, North-South electricity interconnections in Central Eastern and South Eastern Europe, the Baltic region, and to build offshore grid in the North Sea for offshore and onshore renewable integration (European Commission 2014a).

Regarding market structure, we observe that the current oligopoly stands from historical monopolies with few new entrants. Since the first directive of electricity market reform, there were two types of new entrants: The first phase entailed the new entrants around 2005, mostly based on CCGT technology, as it was the most profitable investment given wholesale prices at the time. These new entrants are truly the result of liberalization. The second types are the new entrants based on renewable technologies and driven by FiT and other support schemes. These new entrants are out-of-market new entrants.

However, if we observe company movements between concentration and actor multiplication, we can see that new entrants of medium and small sizes tend to mostly be bought by historical major monopolies, and in particular, cross-border new entrants tend to return to their historical market: French operator EDF had bought a share of EnBW in Germany but ended up selling it to Baden-Wurtenberg; the Italian operator Enel had entered the Romanian market but eventually left it, as

French company Engie did in Slovakia. The main exception is the UK where the major companies, the 'big six', include French, Spanish and German operators. As a result, the EU market is still concentrated around historical operators.

3.3.2 Prices

The EC quarterly report on electricity markets shows that average European whole-sale prices are plunging from levels between 45 and 85 €/MWh in 2008 to levels between 20 and 45 €/MWh in 2016. In particular, in the first quarter of 2016, France, Germany, Benelux all reach prices under 30 €/MWh (European Commission 2016a).

On the other hand, retail prices are rising steadily: in the EU, on average, household electricity prices have risen by 4% a year between 2008 and 2012 (European Commission 2014a, b) which is generally above inflation, mainly because of rising taxes due to renewable support (Eurelectric 2015). Although liberalization was expected to result in cheaper prices, retail consumers do not benefit from low wholesale prices, while power utilities face increasing difficulties to recover their costs.

4 Paradoxes of Electricity Market Liberalization and Climate Policy and Future Trends

4.1 New Stakes in Climate Change and Renewables: Towards a Combined Policy Package

Environmental concerns have grown over the past decades with the creation of Intergovernmental Panel for Climate Change (IPPC) in 1988, the signature of the Kyoto Protocol in 1997, or the Stern Report (Stern 2006, 2007). They leaded Europe to develop an ambitious plan for energy and climate: the Climate and Energy Package defined by the (European Commission 2009a, b, c).[7] The Second Climate and Energy Package with 2030 targets released in 2014 comprised the objectives submitted to the COP21 in 2015: 40% GHG emissions reductions (compared to 1990 levels), 27% renewable energy share in primary energy mix, 27% energy efficiency improvement (European Commission 2015a; UNFCCC 2015).

Regarding renewables, there is a pertinent need for investments coordination through new regulation in all European countries. The share of renewables is indeed growing within all generation mixes across Europe, which raises technical and

---

[7]It plans cutting greenhouse gas emission in 2020 (−20% compared with 1990), increasing energy efficiency (+20% more than *business-as-usual* projections for 2020) and objectives regarding the generation mix (20% renewable energies in the mix).

economic issues such as dealing with intermittency and zero marginal price technology on wholesale markets.

From an economic point a view, it is difficult to find a unified theory allowing to determine optimal pricing and optimal investment amounts when renewables are rising (OECD and Nuclear Energy Agency 2012). This rise indeed makes theories on optimal investment faulty for two reasons. First, incentives such as carbon tax, feed-in tariffs or green certificates distort the data for traditional models based on cost minimization issue from Massé's works. Such models structure costs in terms of fixed costs (investments) and variable costs (operation and maintenance, and fuel). *Ramsey-Boiteux* optimal pricing is based on marginal costs and determines investments from them. However, for unavoidable renewable energies, the variable cost is quasi-zero, resulting in the marginal cost also being zero, which does not allow optimal pricing nor adequate price signal for investments. Besides, the fact that recent renewable technologies (wind, solar) are both unpredictable and intermittent are not yet correctly taken into account in existing models and are still under research. In reality, unpredictability and intermittence of renewables make it necessary to deploy demand response tools in order to compensate for drops in generation like back-up gas-fired plants, and to develop interconnected grids for larger distances to take advantage of the geographical dispersion of renewables. Such heavy investments are only starting to be negotiated or deployed in a few areas of Europe (like Scandinavian countries).

As European climate policy and market liberalization were initially designed as separate policies, it is only very recently that the European objectives have started to evolve towards a combined policy package, in order to ensure compatibility and complementarity of European Commission directives. The European Commission decided upon the creation of an 'Energy Union' as an objective for 2019 (European Commission 2015b; Newbery 2015). It aims at new governance for the energy sector taking climate action into account. As there is an urgent need to reconcile the contradiction between the energy market liberalization policy that focuses on competition and promotes less intervention from the states, and increasing government interventionism for climate policy purposes (Keay and Buchan 2015). The 'Energy Union' includes three kinds of objectives: general geopolitical objectives, climate objectives and market integration objectives. Geopolitical objectives aim at increasing the energy independency of Europe, i.e. diversifying sources, and reducing import dependency of the EU. Climate objectives consist in binding GHG reduction target: -40% in 2030 compared to 1990 levels as in the Second Energy Package released in 2014 and as in the NDCs submitted to the UNFCCC for the COP21(-European Commission 2015a; UNFCCC 2015). It also includes the same EU-wide binding renewable energy target (27%), objectives to improve energy efficiency, and to reform the EU-ETS. In order not only to promote market integration and competition within the market, but also to improve security of supply and reduce CO_2 emissions, the plan for the 'Energy Union' also sees an investment of 647 million € in 'Projects of Common interest' that are mainly energy infrastructure (European Commission 2014).

4.2 The Combined Effect of Renewable Support and Liberalization: Overcapacity and Low Marginal Costs

The combined effects of renewable support and liberalization in Europe cause severe issues for European utilities (Finon and Roques 2013; Newbery 2015; Robinson 2015; Keay 2016; Keay and Buchan). As mentioned earlier, liberalization and climate policy were initially separate packages in EU policy, which means that the EC was creating a liberalized internal market while taking about 20–30% of generation outside the market with FiTs and other support schemes at the same time—renewable generation represented 26% of power generation in 2015 (European Commission 2015c). Support schemes encouraged important investments in renewables ending up in overcapacity given the falling demand, while liberalization induced marginal cost pricing on wholesale markets. Since renewables are fatal energies and have priority access to the network, massive renewable production with very low marginal cost brought wholesale prices down to under 30 €/MWh in early 2016 (Keay and Buchan), reducing power company revenues while increasing some of their costs. In addition, this massive production lowered average wholesale price, peak prices also become lower—yet peak sales are usually a source of profit for baseload and semi-baseload power plants. The overcapacity in Europe resulted in less demand for conventional thermal plants, increasing their average costs. Intermittency causes new system costs, as investments are needed in networks for off-shore wind and in local distribution for distributed solar. Network tariffs rose by 18% between 2008 and 2014 for residential consumers (Eurelectric 2015). As a consequence, energy-only markets are deemed unable to remunerate the fixed costs of power stations; moreover, it induces that there is no exit plan for support schemes. Major power companies are faced with precarious financial realities as the low level of wholesale prices do not allow them to recover variable and even fixed costs, which causes them to suffer from stranded costs and impairments. About 70 GW of coal and gas power plants were shut down between 2010 and 2014, with important acceleration over time: 3 GW per year in 2010 and 2011, 10 GW in 2012, almost 30 GW per year in 2013 and 2014 (Robinson 2015).

Moreover there are additional problems linked with massive renewable investment and the 'missing money' problem for thermal power plants. There is a phenomenon of cannibalism for renewable investment; PV investment especially. For such investment, the additional capacity will produce more electricity at the same peak period (at noon for PV for instance). There is thus less and less value to invest in such additional capacity, as it tends to increase the burden of intermittency-related issues and attendant system costs.

According to (Robinson 2015), in addition to the issues identified above, rising retail prices lower the demand as it encourages savings, efficiency and auto-consumption—a trend that only EV development would change, although high retail prices may discourage EV deployment and thus compromises the decarbonization of transportation—however the first obstacle is the price of the EV itself. Such an interpretation supposes that electricity demand is price-elastic. However, experts tend to say that there is little elasticity for electricity demand, and high retail prices do not affect electricity consumption in the short-term. In the long term however, there could be observable effects such as fuel switch for heating for instance. Besides, this interpretation also supposes that autoconsumption is becoming more and more attractive. Indeed, the literature analyses the potential for autoconsumption causing the death spiral of utilities (Sioshansi 2014), with mentions that rooftop PV had reached grid parity in several countries (such as Italy). But in concrete terms, it is an artificial grid parity, meaning that the cost for auto-consumption is equal to market electricity prices including high taxes, which in no way means that generation costs are equal. Autoconsumption is still assessed to be far from competitive worldwide, even in areas with the highest electricity prices (Khalilpour and Vassallo 2015). It develops unequally within Europe, meeting little success in France or UK, but developing in Scandinavian countries and Germany. For such countries, this aspect could thus potentially become an issue in the future.

In the end, major utilities are facing a critical "missing money" problem, trapped in a vicious circle that does not allow them to pay fixed costs; the system is thus unable to phase out of support schemes and shift to clean energy investments only, and retail customers do not benefit from low wholesale market, as summarized on Fig. 1.

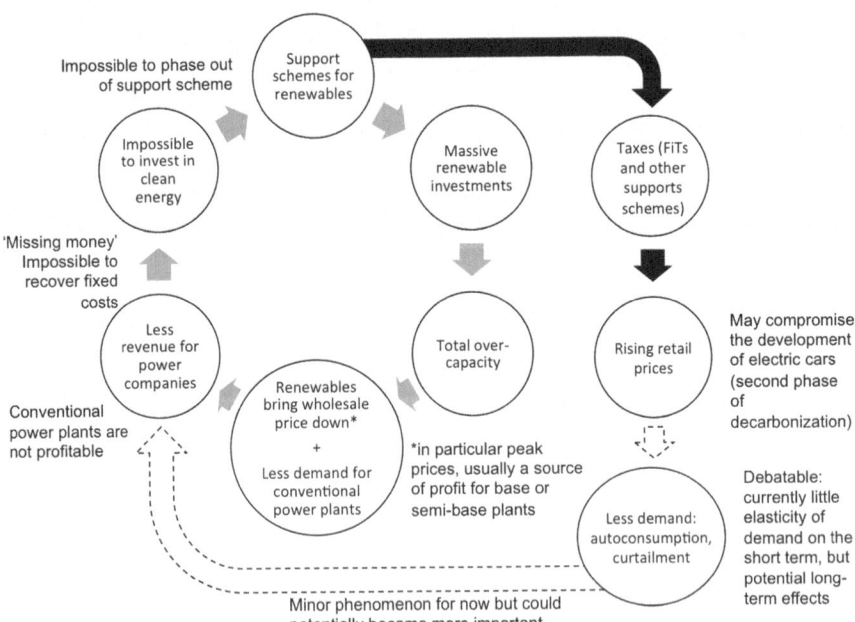

Fig. 1 The combined effect of renewable support and liberalization (author figure)

4.3 Policy Recommendations to Help Investments; Re-regulation or More Liberalization?

In response to these issues, two trends appear as solutions to direct investments: re-regulation ore more liberalization, but with adapted self-balancing mechanisms. Several recommendations are identified for EU from literature review (Finon and Roques 2013; RTE 2014; FTI Intelligence 2015; Keay 2016) and interviews of experts.[8]

4.3.1 Balancing Mechanisms in a Liberalized Market for Self-regulation

These recommendations are heading in a similar direction as the in the sense of the new proposals by the EC called the "Winter Package" (European Commission 2016b).

First, it is commonly accepted that capacity mechanisms are necessary to remunerate the guarantee of generation necessary for peak load investment (Finon and Roques 2013; FTI Intelligence 2015), as it could also allow for the recovering of some of the stranded costs. Second, it is essential to transfer the price signal of wholesale market to retail consumers (industrial, residential), which can be done via smart home system and demand response. The EC however plans to limit such mechanisms to countries or regions where the need for it has been specifically identified.

Moreover, in response to overabundance of allowances in previous phases of the EU-ETS, which was identified as one of the causes for low carbon prices, a stability reserve will be implemented from 2019, allowing for a freeze of a certain proportion of allowance if needed (European Commission 2015d).

Lastly, market-friendly renewable support such as Contract for Difference or Feed-in-Premium instead of Feed-in-Tariffs would allow for exposing producers to market prices in order to prevent overcapacity, while still protecting them from important variations. While Feed-in Tariff provides constant revenue, Feed-in Premium adds a premium to the producer's revenue, which thus follows market variations. Contract for Difference are an alternative between Feed-in Tariff and Feed-in Premium. The revenue of the producer is compensated according to two prices: the reference price and the strike price. The reference price corresponds to the

---

[8]Experts from the following institutes/companies were interviewed: Electricite de France (EDF), Reseau de transport d'electricite (RTE), Enedis (Former ERDF Electricité Réseau Distribution France), World Energy Council, French Energy Council, CREDEN (Montpellier I University), CIRED, CERNA (Mines ParisTech), Institute for Techno-Economics of Energy Systems, CEA Saclay, Laboratory of Industrial Engineering, CentraleSupélec, (Paris-Saclay University), Climate Economics Chair, Paris, Central Research Institute of Electric Power Industry (CRIEPI) of Japan, Institute of Energy Economics of Japan (IEEJ).

average market price on a given period (annual average price, seasonal average price or even hourly price in some cases) and thus follows market though usually with less volatility. When the reference price is below a negotiated 'strike price', the revenue of the producer is compensated to reach the level of the strike price. On the other hand, when the reference price is superior to the strike price, the difference is paid back by the generator. A simplified version of Contract for Difference consists in taking the market price as the reference price.

Although these support schemes are market-friendlier than FiT, they may still at some point be ineffective to prevent overcapacity as they protect the producer anyway.

4.3.2 Re-regulation Recommendations

Re-regulation recommendations are advocated by the power sector and actors wanting to protest national industrial champions. Centralized planning of capacity by the government is one of them, for example through tenders—which would allow competition for the choice of the company for the particular investment—so that the electricity mix is a state decision rather than the result of market price signals. Long-term arrangements are mostly banned by competition regulation, but allowing them in some form would allow revenue guarantee for companies as well as a hedge again price volatility for consumers. There are currently a few financial schemes that are close to such arrangements, such as the Exeltium consortium in France, a group of electro-intensive industrials who benefit from a long-term agreement (24 years) for cheap power supply (Exeltium 2016).

Regarding policy climate instruments, a carbon floor price around 30 € per tCO2 would allow for shifting from coal to gas by changing the merit order as a first step towards a steadily rising carbon price to direct low-carbon investments. It would cut emissions by 15% a year, while increasing the gas share to 40% in European power generation. In the longer term, a 100 € per tCO2 price would allow 30% emissions cuts per year (RTE and ADEME 2016). In practice, the UK has already adopted a carbon floor price starting at a level of 18£ per tCO2 and planning to reach 30£ per tCO2 in 2020 (Government of UK 2014). However adopting such a policy at a national level without guarantee of harmonization within the EU represents a risk in terms of competitiveness. Several countries are considering a carbon floor price such as France and UK. Maintaining relatively high carbon prices raises the issues of the redistribution of CO2 revenue, that could be a source for low-carbon electricity investments (World Bank 2016).

Beyond investments in generation capacities, transmission and distribution management, in terms of both investment and pricing, are foreseen to be the next essential issues (Creti 2016; Galland 2016; Jamme 2016; Meeus 2016; Roques and Verhaeghe 2016; Schwarz 2016; Thouvenin 2016). Both market liberalization and renewable integration require reinforcement of the grid, in transmission as well as distribution, as they will require more and more flexibility. Large-scale wind and upcoming nation-wide capacity mechanisms call for more transportation capacity, while recent massive PV and small renewable investments demand new infrastructure for distribution grids. While the network used to be conceived and designed from power plants to end-use consumers, that is to say from transmission to distribution, from Extra-High Voltage towards Low Voltage, the opposite need revealed itself: from residential customers to main grid. For such 'prosumers' (consumers and producers at the same time), fixed costs of the distribution and transmission network are not rightly remunerated since only net subtracted electricity is billed. A benchmark of current pricing practices for distribution shows that fixed costs are not properly remunerated (compensated for?) in most European countries, which is problematic for investments—except for the Netherlands, Spain and Sweden, where fixed costs account for over 75% of the tariff (Roques and Verhaeghe 2016). There is thus a need to reconsider pricing and reform tariffs (Galland 2016), but also to design incentives such as storage premium so that prosumers can contribute to grid support. Beyond investment and pricing, alternative regulations should be considered to adapt the system to customers' new needs: for instance to design small-scale capacity mechanisms at city levels or neighborhood levels, managed by distribution operators rather than nation-wide capacity mechanisms managed by transportation operators (Creti 2016; Meeus 2016).

5 Conclusion

The study allows for mapping out past and current issues in Europe for investments in electricity capacities. Throughout this chapter, we can see that the drivers for the investment decision in electric power capacities have been getting more and more complex as new imperatives piled up, from the primary concern of energy security to current issues of fair competition and environment and climate protection. As a result, both regulatory risks and market risks are so high that power utilities are unable to invest while stranded costs appear; there are very few actors left, who are in a position to make the necessary low-carbon investments.

This chapter concludes with the level of government intervention that is desirable in a post-reform electricity sector and sorts out existing policy instruments to reconcile requirements for a liberalized market and an effective climate policy.

Acknowledgements The authors would like to address special thanks to the interviewed experts for their time and precious insights: C. Bonnery (Enedis), A. Creti (Climate Economics Chair, Paris Dauphine), J.-G. Devezeaux and I-tese team, D. Finon (CIRED), L. Joudon (EDF), F. Lévêque (CERNA), J.-E. Moncomble (French Energy Council), J. Percebois (CREDEN, Climate Economics Chair), Y. Perez (CentraleSupelec), C. de Perthuis (Climate Economics Chair), D. Sire (World Energy Council), B. Solier (Climate Economics Chair), T. Veyrenc (RTE), K. Akimoto and Y. Nakagami (RITE).

References

Allais, M. (1943). *A la recherche d'une discipline économique*. Industria: Impr.

Arrow, K. J. (1965). *Aspects of the theory of risk-bearing*. Helsinki: Yrjö Jahnssonin Säätiö.

Asociasion espanola de la industria Electrica. (2013). *UNESA – Historia*. In: UNESA. Accessed July 31, 2013, from http://www.unesa.es/que-es-unesa/historia

Averch, H., & Johnson, L. (1962). Behavior of the firm under regulatory constraints. *American Economic Review, 52*(5), 1052–1069.

Baumol, W. J. (1977). On the proper cost tests for natural monopoly in a multiproduct industry. *American Economic Review, 67*, 809–822.

Beltran, A., & Bungener, M. (1987). Itinéraire d'un ingénieur. *Vingtième Siècle Rev Hist, 15*, 59–68. https://doi.org/10.2307/3769628.

Boiteux, M. (1956). Sur la gestion des monopoles publics astreints à l'équilibre budgétaire.

Buchanan, J. M. (1975). *The limits of liberty: Between anarchy and leviathan*. Chicago: University of Chicago Press.

Chick, M. C. (2007). *Electricity and energy policy in Britain, France and the United States Since 1945*. Cheltenham: Edward Elgar Publishing.

Creti, A. (2016). *Economic criteria for tarification*. Synthesis report of the Conference on Grid Tarification (AEE, CGEMPE, CGR, CEEM).

Denant-Boèmont, L., & Raux, C. (1998). Vers un renouveau des méthodes du calcul économique public? *Metro, 106–107*, 31–38.

Department of Energy & Climate Change, Prime Minister's Office. (2013). *Initial agreement reached on new nuclear power station at Hinkley*. Press releases – GOV.UK. Accessed November 6, 2013, from https://www.gov.uk/government/news/initial-agreement-reached-on-new-nuclear-power-station-at-hinkley

Enerdata Natural Gas Production. (2012). *Statistics about gas natural production*. Accessed November 3, 2013, from http://yearbook.enerdata.net/#world-natural-gas-production.html

Eurelectric. (2015). *Power statistics and trends*.

European Commission. (2009a). Directive 2009/28/EC of the European Parliament and of the Council of 23 April 2009 on the promotion of the use of energy from renewable sources and amending and subsequently repealing Directives 2001/77/EC and 2003/30/EC.

European Commission. (2009b). Directive 2009/29/EC of the European Parliament and of the Council of 23 April 2009 amending Directive 2003/87/EC so as to improve and extend the greenhouse gas emission allowance trading scheme of the Community.

European Commission. (2009c). Directive 2009/31/EC of the European Parliament and of the Council of 23 April 2009 on the geological storage of carbon dioxide and amending Council Directive 85/337/EEC, European Parliament and Council Directives 2000/60/EC, 2001/80/EC, 2004/35/EC, 2006/12/EC, 2008/1/EC and Regulation (EC) No 1013/2006.

European Commission. (2014a). *Single market progress report*.

European Commission. (2014b). *Energy prices and costs in Europe*.

European Commission. (2015a). *Intended nationally determined contribution of the EU and its member states*.

European Commission. (2015b). *A framework strategy for a resilient energy union with a forward-looking climate change policy*.

European Commission. (2015c). *Renewable energy progress report*.

European Commission. (2015d). *State of the energy union*. Climate action progress report, including the report on the functioning of the European carbon market and the report on the review of Directive 2009/31/EC on the geological storage of carbon dioxide.

European Commission. (2016a). *Quarterly report on European electricity markets*.

European Commission. (2016b). *Clean energy for all Europeans*.

Exeltium the project. (2016). *Exeltium*. Accessed August 18, 2016, from http://www.exeltium.com/project/?lang=en

Finon, D., & Roques, F. (2013). European electricity market reforms: The "Visible Hand" of public coordination. *Econ Energy Environ Policy*. https://doi.org/10.5547/2160-5890.2.2.6.

Francony, M. (1979). Theory and practice of marginal cost pricing: The experience of "electricite De France". *Ann Public Coop Econ, 50*, 9–36. https://doi.org/10.1111/j.1467-8292.1979. tb00838.x.

Friedman, M., & Savage, L. J. (1948). The utility analysis of choices involving risk. *Journal of Political Economy, 56*, 279–304.

Galland, J. B. (2016). *The distributor's point of view*. Synthesis report of the Conference on Grid Tarification (AEE, CGEMPE, CGR, CEEM).

Glachant, J. M. (2000). Les pays d'Europe peuvent-ils reproduire la réforme électrique de l'Angleterre? Une analyse institutionnelle comparative. *Économie Prévision, 145*, 157–168. https://doi.org/10.3406/ecop.2000.6121.

Government of UK. (2014). *Carbon price floor: Reform*. Business tax – Policy paper.

Grand, E., & Veyrenc, T. (2011). *L'Europe de l'électricité et du gaz*. Econometrica.

Henry, C. (1974). Investment decisions under uncertainty: The "Irreversibility Effect". *American Economic Review, 64*, 1006–1012.

Hicks, J. R. (1939). The foundations of welfare economics. *The Econometrics Journal, 49*, 696. https://doi.org/10.2307/2225023.

Ibeas Cubillo, D. (2011). *Review of the history of the electric supply in Spain from the beginning up to now*. Bachelor Thesis.

FTI Intelligence. (2015). *Toward the target model 2.0*.

Jamme, D. (2016). *Issues in the tarification of electricity grid (TURPE)*. Synthesis report of the Conference on Grid Tarification (AEE, CGEMPE, CGR, CEEM).

Kahn, A. E., & Eads, G. (1971). A. E. Kahn: The economics of regulation. *Bell J Econ Manag Sci, 2*, 678. https://doi.org/10.2307/3003012.

Keay, M. (2016). *Electricity markets are broken: Can they be fixed?* Oxford: Oxford Institute for Energy Studies.

Keay, M., & Buchan, D. (2015). *Europe's energy union – A problem of governance*. Oxford: Oxford Institute for Energy Studies.

Khalilpour, R., & Vassallo, A. (2015). Leaving the grid: An ambition or a real choice? *Energy Policy, 82*, 207–221. https://doi.org/10.1016/j.enpol.2015.03.005.

Laffont, J. J., & Tirole, J. (1993). *A Theory of Incentives in Procurement and Regulation*. MIT Press.

Leibenstein, H. (1966). Allocative efficiency vs. "X-Efficiency". *American Economic Review, 56*, 392–415.

Massé, P. (1953). Les investissements électriques. *Revue Statistique Appliquée, 1*, 119–129.

Meeus, L. (2016). *Evolution of regulation for new services and uses*. Synthesis report of the Conference on Grid Tarification (AEE, CGEMPE, CGR, CEEM).

Mills, S. C., & Williams, R. (1986). *Public acceptance of new technologies: An international review*. London: Croom Helm.

National Coal Mining Museum Statistics in Coal Mining. (2013). Accessed November 3, 2013, from http://www.google.fr/url? sa=t&rct=j&q=&esrc=s&source=web&cd=1&ved=0CDIQFjAA&url=http%3A%2F% 2Fwww.ncm.org.uk%2Fdocs%2Fcollections-documents%2Fstatistics-in-mining.pdf% 3Fsfvrsn% 3D2&ei=SoB2Uqn5IInB0gW4toCYBA&usg=AFQjCNGcA1nloiJbexQz5VpIdAM25RPeK-Q&sig2=d3YqEVXGA42W1a9npPeiTQ&bvm=bv.55819444,d.d2k&cad=rja.

Neumann, J. V., & Morgenstern, O. (1944). *Theory of games and economic behavior*. Princeton, NJ: Princeton University Press.

Newbery, D. (1997). Privatisation and liberalisation of network utilities. *European Economic Review, 41*, 357–383. https://doi.org/10.1016/S0014-2921(97)00010-X.

Newbery, M. (2015). *European energy handbook 2015: A survey of current issues in the European energy sector*. London: Herbert Smith Freehills.

OECD, Nuclear Energy Agency. (2012). *Nuclear energy and renewables*. Paris: Organisation for Economic Co-operation and Development.

Office statistique des Communautés européennes. (2016). *The production of coal and steel in Europe (1936–1958)*.

Peltzman, S. (1976). Toward a more general theory of regulation. *Journal of Law Economics, 19*, 211. https://doi.org/10.1086/466865.

Perrot, A. (2002). Les frontières entre régulation sectorielle et politique de la concurrence. *Revue Française d'Économie, 16*, 81–112. https://doi.org/10.3406/rfeco.2002.1522.

Pigou, A. C. (1924). *The economics of welfare*. Piscataway, NJ: Transaction Publishers.

Robinson, D. (2015). *The scissors effect: How structural trends and government intervention are damaging major European electricity companies and affecting consumers*. Oxford: Oxford Institute for Energy Studies.

Roques, F., & Verhaeghe, C. (2016). *Benchmark of tarification for distribution grid*. Synthesis report of the Conference on Grid Tarification (AEE, CGEMPE, CGR, CEEM).

RTE. (2014). *French capacity market – Report accompanying the draft rules*.

RTE, ADEME. (2016). *CO2 price effect on power generation mix* (Effets prix du CO2 sur mix électrique).

Samuelson, P. A. (1943). *Foundations of economic analysis*. Cambridge, MA: Harvard University Press.

Schwarz, V. (2016). *Tariffs as energy policy instruments*. Synthesis report of the Conference on Grid Tarification (AEE, CGEMPE, CGR, CEEM).

Sharkey, W., & Reid, G. C. (1983). The theory of natural monopoly. *Economic Journal, 93*, 929. https://doi.org/10.2307/2232765.

Sioshansi, F. P. (2014). Introduction: The rise of decentralized energy. In F. P. Sioshansi (Ed.), *Distributed generation and its implications for the utility industry* (pp. xxxiii–xxxili). Boston: Academic Press.

Stern, N. (2006). The stern review on the economic effects of climate change. *Population and Development Review, 32*, 793–798. https://doi.org/10.1111/j.1728-4457.2006.00153.x.

Stern, N. (2007). *Stern review report on the economics of climate change*. HM Treasury.

Stiglitz, J. E. (1976). Monopoly and the rate of extraction of exhaustible resources. *American Economic Review, 66*, 655–661.

Thouvenin, V. (2016). *The transmission operator's point of view*. Synthesis report of the Conference on Grid Tarification (AEE, CGEMPE, CGR, CEEM).

UNFCCC. (2015). *INDCs as communicated by parties*.

Weisbrod, B. A. (1964). Collective-consumption services of individual-consumption goods. *Quarterly Journal of Economics, 78*, 471–477. https://doi.org/10.2307/1879478.

World Bank. (2016). *State and trends of carbon pricing 2016*. The World Bank.

A Prospective Analysis of CO_2 Emissions for Electric Vehicles and the Energy Sectors in China, France and the US (2010–2050)

Wenhui Tian and Pascal da Costa

Abstract Within the landscape of global warming and energy transition, many countries have announced nationally aligned contributions in reducing their CO_2 emissions (COP21 and 22, in 2015 and 2016 respectively). With the aim of evaluating the maturing and the success of these targets, technology roadmaps are necessary and serve a twofold function in the evaluative process. They serve as points of comparisons between each other and they are yardsticks by which to measure change for the 2050 horizon.

In this chapter, technology roadmaps are studied for three representative countries: China, France and the United States of America. The roadmaps cover the sectors responsible for the greatest part of CO_2 emissions, i.e. the power, transport, residence and industry sectors. They also cover the impact of the main technologies, i.e. carbon capture and storage, energy efficiency and electric vehicles. This chapter thus assesses the future of energy trends and especially shows that the deployment of electric vehicles shall prove crucial for reaching the commitments towards contributions at national levels.

Keywords Energy transition · Technology roadmaps · Sectoral emission modeling · STIRPAT model · Support vector regression

1 Introduction

Numerous countries have submitted to the United Nations Framework Convention on Climate Change for the COP21 in Paris on 2016 December (UNFCCC 2015), their nationally determined contributions in reducing the emissions of CO_2 that is the most important Greenhouse Gas (GHG) (IPCC 2013); a marked commitment

Parts of this chapter were published in Open Access in 2015 on https://hal.archives-ouvertes.fr/hal-01026302v3/document (Da Costa, P., Tian, W. (2015). A Sectoral Prospective Analysis of CO_2 Emissions in China, USA and France, 2010-2050, HAL w.p.).

W. Tian · P. da Costa (✉)
Laboratoire Genie Industriel, CentraleSupelec, Université Paris-Saclay, Gif-sur-Yvette, France
e-mail: pascal.da-costa@centralesupelec.fr

towards reducing global warming. The objective of this chapter is to assess these policy targets by evaluating substitutable technology roadmaps within the period of energy transition, from 2010 to the year 2050. This period often perceived as a turning point in energy use patterns worldwide forced by the decline in hydrocarbon extraction. In this chapter we will simulate a flexible modeling framework in order the better understand both the future trends in energy and the changes to be made compared to today.

Several families of climate change economic models already work on technology roadmaps. They co-exist with huge differences of decompositions, be they at sectorial, regional or fiscal levels. These differences exist also with the theories used, for instance with endogenous or exogenous growth, the different market structures and so on, within long or mid-term perspectives (Chen 2005; Klaassen and Riahi 2007; Saveyn et al. 2012). Therefore the mechanisms and assumptions of these models are often opposed which make it difficult to compare their results and well understand the numerous differences in predictions (see previous work of Boulanger and Bréchet 2005 or that of Akimoto 2016 about the comparisons of climate change economics models: DN21+, WITCH, AIM, etc.). These models (DN21+, MARKAL, WITCH, AIM, NEMS, etc.) finally require large amounts of exogenous input and have complex structures with fairly limited access.

In this chapter, we propose a less complex (less data required, simple framework) but complementary approach, by taking into account the main energy sectors and energy-related technologies. Our model is based on the IPAT or Kaya identity (Kaya 1989) that plays a core role in the development of future emissions scenarios in some of the IPCC reports (IPCC 2007, 2013) and the IAE studies (IEA 2008).

In this chapter, the transport sector will be studied by evaluating the use of electric vehicles, since this sector represents a huge source of reduction of CO_2 emissions in the future and the electric vehicles technology could be developed quite rapidly. Then the power sector will be depicted trough the energy mix, with the penetration of renewable power, and the potential technology of Carbon Capture and Storage (CCS). Note the use of electric vehicles would undoubtedly have a significant impact on the output of the electricity sector, which made them central to our analysis. In the future, this would lead to a new hybrid energy system, with the connection of power and transport systems. Finally the residence and the industry sectors will also be considered in the model through the improvement of energy efficiency.

The model will be applied to three types of countries that may be considered to be representative of numerous other countries in the world: China (CN), as a fast-emerging economy with increasing energy consumption requirements, and the largest emitter of CO_2; France (FR), a well-developed economy with relatively low CO_2 emissions; and the United States of America (US), the largest economy and a major source of CO_2 emissions.

Within this framework, numerous available solutions for technology pathways can be generated with the model, offering the policy makers choices in technology transitions. The results of the model show that complete changes in the energy structure in all sectors are necessary to achieve governmental reductions in CO_2. We will show that these changes are nonetheless contrasting between countries and that several alternative solutions in terms of technological roadmaps are available to them. In addition to the substitutable pool of technology roadmaps, we will provide two more prospective scenarios: first, under the assumption of 'balanced technology development' across sectors that refers to the same improvement of each technology in reducing CO_2 emissions; and second, under the assumption of 'least changed energy mix' that refers to the minimization of the difference between the energy mix in 2050 and 2010.

The remainder of this chapter is organized as follows. Section 2 introduces our model. Section 3 presents the data for CO_2 emission objectives. Section 4 then explains the results obtained from the model according to different countries. Conclusions are drawn in Sect. 5.

2 The Model of CO_2 Emissions for Energy Sectors

Our model is proposed to evaluate the feasibility of CO_2 mitigation targets with respect to the population, the economy and CO_2 emissions, between 2010 and 2050, within the overall prospect of energy technology transition. It focuses on three types of sectors: power, transport, residence and industry. For the power sector, electricity can be produced from different energy sources. The sources that produce CO_2 emissions are mainly fossil fuels: coal, oil and gas. The clean energies are renewable energies and nuclear energy. In the transport sector, we focus on road transport that generally accounts for more than 80% of transport sector emissions.

Thus the total of CO_2 emissions is the sum of the emissions from power, transport and other sectors: $E_{(t)} = E_{P(t)} + E_{T(t)} + E_{R(t)}$, where $E_{(t)}$ is the total CO_2 emissions from fuel combustions in year t, and $E_{P(t)}$, $E_{T(t)}$ and $E_{R(t)}$ are CO_2 emissions in the corresponding three sectors.[1]

2.1 Power-Generation Sector

In the power sector, we employ the IPAT identity to study the driving forces of CO_2 emissions in producing electricity from fuel combustion. IPAT was developed as a general approach for discussing the driving forces behind environmental impacts, which relates impact (I) to population (P) multiplied by affluence (A) and technology (T)

IPAT was later developed into Kaya identity (Kaya 1989). According to this identity, emissions can be decomposed into the product of three basic factors, carbon intensity of energy, energy intensity and affluence:

---

[1]The details of our model can be found in Appendix 1.

$$CO_2 \; emissions = Population * \frac{GDP}{Population} * \frac{Energy \; consumption}{GDP}$$

$$* \frac{CO_2 \; emissions}{Energy \; consumption} \tag{1}$$

Equation (1) decomposes the CO_2 emissions in the power sector into the product of the output of electricity and the technology (i.e. the emission intensity of production).

Fossil fuels are the main sources of CO_2 emissions in the power sector. Analytically its emissions in year t are divided into three categories as follows:

$$E_{P(t)} = \left(Q_t * \sum x_{i,t} * e_{i,t} \right) * \epsilon(ccs) \tag{2}$$

where Q_t is electricity output, x_i the three main fuels: coal, oil and natural gas; equally e_i is the CO_2 emissions from using coal, oil and gas respectively, and $\epsilon(ccs)$ the dummy variable to which we shall return.

2.1.1 Energy Mix

Energy mixes vary considerably from one country to another.

China has abundant coal reserves, while its oil, natural gas and other fossil energy resources are limited. Coal is currently the dominant power fuel. At the end of 2010, thermal power accounted for 73.4% of total power-generation capacity (IEA 2011).

France is one of the least CO_2 intensive industrialized economies, thanks to the substantial role of nuclear power and the existence of higher gasoline taxes with incentive impacts. In 2009, nuclear power accounted for 76.24% of France's electricity generation. CO_2 emissions have been declining since 2005 from an already relatively low base (IEA 2011).

The US depends on fossil fuels for almost all its energy supply. Natural gas use is growing fast, particularly for power generation, where it has now overtaken nuclear to become the second most important power-generation fuel. Coal is also an important fuel in the US, accounting for 45% of the country's electricity generation (IEA 2011).

As a result of the different energy mix in the power sector, CO_2 emissions per kWh (the emission intensity of production) from electricity generation vary greatly across countries. Figure 1 shows CO_2 emissions per KWh in France are only 12% of the level in China and 20% of the level in the US over 1990–2010, as coal plays a dominant role in China and the US, while nuclear power plants in France.

The emission intensity of the production of each fuel are manifold between countries according to the different types of energy and technology levels, as shown in the Table 1. As the emission intensities of production of fuels are the lowest in Europe, we adopt the emission intensities of production in 2010 of Europe

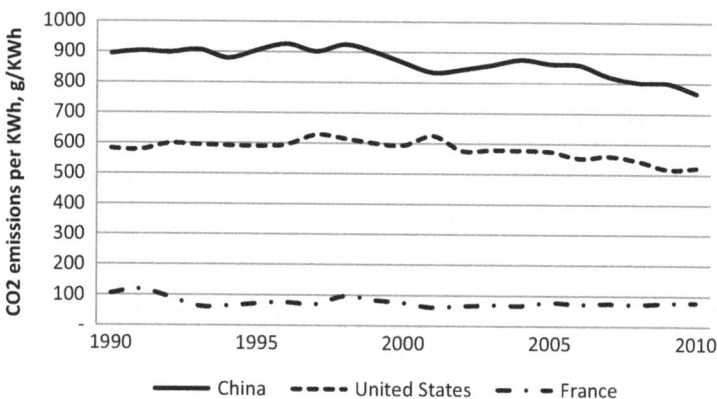

Fig. 1 CO$_2$ emissions per KWh from electricity generation

Table 1 CO$_2$ emission intensities of production of fossil fuels (IEA 2012)

Kg/KWh	Coal	Oil	Gas
China (2010)	0.967	1.044	0.507
France (2010)	0.949	0.766	0.520
US (2010)	0.907	0.711	0.405
Europe (2010)	0.8	0.4	0.2

as the intensities of production for the three countries in 2050, which are 0.8 kg/kWh for coal, 0.4 kg/kWh for oil, and 0.2 kg/kWh for gas.

2.1.2 Electricity Output

As for the electricity production, we project from historical data between 1971 and 2010. Support Vector Regressions (SVR) (Cortes and Vapnik 1995; Gao et al. 2002) are used in order to simulate the projection of electricity output.[2] SVR lends itself well to small databases and has successfully been tested to solve forecasting problems in many fields such as financial time series forecasting (Cao 2003) and electric load forecasting (Hong 2010; Wang et al. 2009). Based on these applications, we have used SVR to make predictions for electricity production and pollution intensity.

The electricity-production simulation results are based on data of 1971–2010 (IEA 2012). The results of simulation show Chinese electricity output will be 10248 tWh in 2050, a 2.43-fold rise over 2010. In France it will be 539 tWh in 2050 (a 4% fall from 2010), and 4785 tWh (a 10% rise over 2010) in the US.

---

[2]The details of parameter settings and electricity outputs can be found in Appendix 2.

2.1.3 Carbon dioxide Capture and Storage Technology

CCS is considered as one of the potential options for end-of-pipe reduction in atmospheric CO_2 emissions from human activities. CCS involves the use of technology, first to collect and concentrate the CO_2 produced in energy-related sources, transport it to a suitable storage location, and then store it away from the atmosphere for a long period of time. CCS would thus allow fossil fuels to be used with low GHG emission.

In fact the potential of CCS technology is at the root related to the CO_2 sequestration potential into aquifer which is unequally distributed among countries (RITE 2015). Today CCS has not been used in large-scale power plants, so there is relatively little experience with the combination of CO_2 capture, transport and storage in a fully integrated CCS system. The wide range of abatement costs of CCS systems is due to the variability of sit-specific factors, the type and costs of fuel used, the required distances, terrains and quantities involved in CO_2 transport, and the type and characteristics of CO_2 storage. In most CCS systems, the cost of capture is the largest cost component, in the range of US\$15–75/tCO_2 net captured from a coal or gas fired power plant. The cost of transportation is between US\$1/tCO_2 and US\$8/tCO_2. The cost of storage is US\$0.5–8/tCO_2 for geological storage and US \$5–30/tCO_2 for ocean storage (IPCC 2005, 2013).

According to prospective studies of (IPCC 2005, 2013), if a power plant is equipped with CCS technology, about 90% of the CO_2 emissions could be captured and stocked. We assume that CCS technology is developed and widely disseminated in the year 2050 and the dummy variable ϵ(ccs) in the Eq. (2) is equal to:

$$\epsilon(ccs) = \begin{cases} 0.1, & 90\%\text{emissions be absorbed with CCS} \\ 1, & \text{no emissions be absorbed without CCS} \end{cases}$$

2.2 Transport Sector

The power generation sector makes the most important contribution to global CO_2 emissions, and the second most important emitter is the transport sector. The transport sector was responsible for approximately 23% of the global energy-related CO_2 emissions in 2010. Note 72% of CO_2 emissions come from road transport (IEA 2011).

2.2.1 Road Transport Vehicles

Reducing global transport GHG emissions will be challenging due to the continuing growth of passenger and freight activities in some areas. The transportation sector accounted for over 40% of oil demand in the world in 2010. Oil use will become increasingly concentrated in the transportation sector, reaching 65% of total oil demand in 2035 (IEA 2011). Thus automobiles with clean energy sources are encouraged to replace the traditional gasoline and diesel ones.

Hybrid vehicles and electric vehicles are two emerging technologies that manufacturers are increasingly turning towards, especially for the electric vehicles. Hybrid vehicles (conventional hybrids, plug-in hybrid electric vehicles) combine both an electric motor and a gasoline engine. Electric vehicles (plug-in, battery and fuel cell electric vehicles) use an electric-only motor but with different energy storage systems. Electric vehicles have no direct tailpipe emissions; the indirect emissions come from charging the vehicle's battery with grid electricity generated by fossil-fuel-powered power plants. Thus electric vehicles have a CO_2 reduction cost that is highly correlated to carbon intensity of electricity generation. However, with the transformation of the power sector, indirect emission will reduce in the long term. In this context we choose electric vehicles as the option for the technology transition in transport sector modeling. Along with the advantages of electric vehicles, there are barriers for the adoption, such as high battery costs, willingness of consumers, charging facility, etc. The penetration of the market and technology advancement need the encouragement of governments.

2.2.2 Road Transport Emission Across the Three Countries

The transport sector is responsible for the largest share of CO_2 emissions in France (over one third of emissions in 2010), with road transport accounting for 96% of transport emissions. Thanks to its low-cost and low-carbon electricity supply, France has been able to reduce transport emissions by focusing on electricity-based technologies, such as high-speed rail and electric vehicles. There are currently about 30,000 electric vehicles in France: only 0.08% of all the vehicles. The French energy transition law adopted in 2015, announced that the bonus for changing to electric vehicles can be accumulated up to 10,000 euros, and the government will install charging stations all over France, with the objective of a total of 7 million in 2030.

In 2010, the US had the largest number of vehicles out of any of the countries in the world (254 million), with transport accounting for 30% of CO_2 emissions, and road emission responsible for 86.4%. In 2009, the US President pledged US$2.4 billion in federal grants to support the development of next-generation electric vehicles and batteries. As part of the American Recovery and Reinvestment Act, the Department of Energy announced the release of two competitive solicitations for up to $2 billion in federal funding for competitively awarded cost-shared agreements for manufacturing advanced batteries and related drive components as well as up to $400 million for transportation electrification demonstration and deployment projects. This initiative aimed to help meet the President's goal of putting one million plug-in electric vehicles on the road by 2015. In 2014, nearly 120,000 electric vehicles were sold in the US.

In China, transport accounts for only 7% of total emissions in 2010. With a growth rate of 11% of the number of vehicles in 2010, transport—road transport especially, will be increasingly important for future CO_2 emissions. Thus it proves critical for China to develop electric vehicles. 2012 saw 4400 electric vehicles in circulation. In order to encourage the consumers, 5 billion Yuan was allocated as the total allowance for the purchasing of electric vehicles from 2009, and 5 million electric vehicles in 2020.

With current technology, an electric vehicle consumes 0.01 KWh/km to 0.03 KWh/km. Here we employ the mean value of 0.02 KWh/km, that is 0.73 MWh/year. (shown in Eq. (3) below) that makes a notable contribution to total electricity output. The CO_2 emissions of the transport sector are calculated as:

$$E_{T(t)} = \frac{E_{road(t)}}{\alpha_{road}} = \frac{E_{road(2010)} * (1 + \gamma)^t * (1 - y_t)}{\alpha_{road}} \tag{3}$$

where $E_{T(t)}$ are CO_2 emissions in the transport sector, $E_{road(t)}$ are CO_2 emissions from road transport, is the vehicle growth rate, y_t is the proportion of hybrid vehicles in the vehicle stock, and α_{road} is the share of road transport in the emissions of the transport sector. The baseline emissions in the road transport will increase from 400 mt in 2010 to 1968 mt in 2050 in China, due to the fast growth of car numbers. The baseline in France will increase from 118 mt to 198 mt; from 1400 mt to 2170 mt in the US.

The use of hybrid vehicles will definitely increase electricity production as follows:

$$E_P^{tr} = 0.73 * y_t * N_{(t)}$$

where $N_{(t)}$ is the stock of vehicles. Total electricity output is therefore $E_{P(t)} + E_P^{tr}$. For the number of vehicles in 2050, we assume that it will keep increasing at this growth rate in 2010, of about 1% in France and in the US, as their car number growth was at a stable rate.[3] However, because the car numbers were increasing fast in the past few years in China, we assume the cars numbers will increase first at a fast rate as in 2010 at 10%, and then this growth rate will progressively decrease to 1% in 2050.[4]

The numbers of vehicles in 2050 in the three countries in our study are shown in Table 2. In our assumption, vehicles in China will increase much more than the other two countries, from 114 million to 560 million in 2050: it is at the same level as in 6DS scenario (baseline scenario) in (IEA 2014). The car numbers will rise from

---

[3]The projections of cars number in France and the US are more optimist than those of the (IEA 2014). The car number projections in different studies can be controversial in terms of various assumptions. For example the personal cars number in 2050 are projected to be about half of that in 2010 according to the projection of (Alazard-Toux et al. 2014). In this chapter, we project evolution of the car numbers in countries following their historical growth trends without involving other parameters in order to make a simplified and clear assumption.

[4]The projection of cars number in China in 2050 is at the same level than (IEA 2014) baseline.

Table 2 Assumptions for number of vehicles in 2050

	Number of vehicles in 2010 (million)	Number of vehicles in 2050 (million)	Number of vehicles per person in 2010	Number of vehicles per person in 2050
China	114	560	0.085	0.4
France	38	63	0.6	0.9
US	268	393	0.86	0.98

38 million to 63 million in France, and from 269 million to 393 million in the US, which means nearly one vehicle per person.

2.3 Domestic and Industrial Sector

Improving energy efficiency is a key for reducing GHG emissions in the domestic and industrial sectors. However, as the domestic and industrial sectors are not the key sectors to be studied in this work, we employ the overall improvement of energy efficiency in these sectors instead of assessing the detailed energy efficiency technologies.

2.3.1 Energy Efficiency Related Technologies

The energy use and related emissions in the domestic sector will increase, especially in developing countries, with the increasing need for adequate housing, electricity and improved cooking facilities. For the industrial sector, despite its declining share in global GDP, the GHG emissions from the industrial sector keep increasing. In 2010, domestic CO$_2$ emissions accounted for 22.4% of those in the other sectors in the US; in France and China, this figure was 31.8% and 9.6% respectively. Improving energy efficiency can reduce domestic and industrial energy consumption.

Energy efficiency as a general notion involves consuming less energy in providing the same service. Many potential technologies are available for improving the energy efficiency. For example more efficient appliances, smart meters and grids, fuel-switching to low-carbon fuels such as electricity or biomass, more efficient insulation in the buildings, and so on. As to the industrial sector, energy efficiency involves fuel switching to low-carbon fuels, efficient process heating systems, materials recycling, etc. For developing countries, there are still many energy efficiency options both for process and system-wide technologies and measures.

CO$_2$ emissions in the rest sectors are presented as follows:

$$E_{R(t)} = \frac{E_{RI(t)}}{\beta} = \frac{(1-e) * E_{RI(baseline)}}{\beta}$$

where $E_{R(t)}$ is CO_2 emissions from the other sectors, $E_{RI(t)}$ is domestic CO_2 emission, e represents the improvement of domestic energy efficiency, $E_{RI(baseline)}$ is domestic CO_2 emission without taking energy efficiency into account, and β is the domestic share in other sector CO_2 emissions. The baseline domestic CO_2 emissions in China will increase from 303 mt in 2010 to 458 mt in 2050, and they will increase from 322 mt to 430 mt in the US. However the baseline emissions in the domestic sector will be reduced from 57 mt to 39 mt in France, because of the decreasing trend of CO_2 emissions in the past few years.

3   The Data

Now we present the emission reduction efforts of nationally determined contributions annouced in the Cancun Agreements. Note that 119 countries submitted on October 1st, 2015, representing 88% of global emissions in 2010. The US House of Representatives passed the Clean Energy and Security Act, that aimed to reduce 17% of their CO_2 emissions below the 2005 level in 2020,[5] and 83% in 2050 (Waxman and Markey 2009): this means that their emissions are expected to be reduced to 981 mt in 2050.

China promised to reduce its CO_2 intensity by 40–45% in 2020[6] (ERI 2009) compared to 2005, and this objective is extendable to 85–90% in 2050. In this work, we adopt the reduction of CO_2 intensity by 90% in 2050, that means the expected emissions are 5 259 mt in 2050, with the baseline scenario of GDP assumption.[7]

The French government announced a reduction of CO_2 emissions by 75%[8] ("Facteur 4") in 2050 compared to level in 1990 (ADEME 2014): this means that the CO_2 emission are expected to be 89 mt in 2050.

As these objectives are brought back to the same time horizon, the proportions of CO_2 emissions by governmental targets in 2050 relative to 2010 will be 72% of that in 2010 in China, 25% in France and 18% in the US (see post-2020 nationally determined contributions in Akimoto 2016).

---

[5]In 2013, the GHG emissions were 9% below 2005 level, according to the "U.S. Greenhouse Gas Inventory Report: 1990–2013".

[6]In 2013, the CO_2 intensity had been decreased by 28.5% compared to 2005. According to the "Plan for the Climate Change (2014–2020)" released in september in 2014 by the Chinese government, the objective of reducing CO_2 intensity in 2020 was not changed.

[7]This CO_2 emission is calculated with the baseline scenario of GDP according (HSBC 2011). Note the GDP using Purchasing Power Parities in China will be \$57,784 billion in 2050, about six times of the 2010 level.

[8]In 2012, the CO_2 emissions from the fuel combustions in France were 5.4% less than its 1990 level, according to (MEDDE and CDC Climat Recherche 2015).

4 The Results for Technology Roadmaps

The reduction of CO_2 emissions in the model are decomposed into the reductions of sectoral emissions, meaning that the reductions of sectors are good substitutes. Thus there exists infinite technology pathways in meeting the scenario objective. That is why we plan to present a solution pool of technology roadmaps only based on technology development. The energy related technologies are: the share of coal and gas in the power sector[9]; the share of electric vehicles; and the improvement of energy efficiency in the residence and industry.

Clearly the share of coal and gas in the power sector are between 0 and 100%, with their sum inferior to 100%. The share of electric vehicles in all vehicles in road transport is in the interval of [0%, 100%], and the improvement of energy efficiency in the residence and industry sector is in the interval of [0%, 100%]. In order to avoid the numberless solutions, we make the following assumptions: *i*) the shares of electric vehicles are set from 0% to 100% with the interval of 20%; *ii*) the improvements of energy efficiency in the residence and industry sector are set from 0% to 80% with the interval of 20%.

Then we discuss two technology roadmaps based on two criteria. One is based on the assumption that technology development across sectors is homogenous. The other is based on the preference of the most use of the energy sources of each country, which means to keep the change of energy mix in the power sector as little as possible.

4.1 Technology Roadmaps in China

We now present the solution pool of the technology roadmaps for China.

4.1.1 Technology Solution Pool in China

In China in 2010, half the CO_2 emissions from fuel combustion came from the power sector, with 78.7% electricity production from the combustion of coal. Thus, the reduction of emissions in the power sector is indispensable. The transport sector contributed only 7% of CO_2 emissions in 2050, but its reduction of emissions can not be ignored due to the fast growth of cars.

According to the governmental target, CO_2 intensity is projected to be reduced by 90% in 2050 with respect to 2005. This means that CO_2 emissions will be reduced by 28% compared to the level in 2010. Different technology pathways are shown in

---

[9]The share of oil is not presented because negligible compared to that of coal and gas, normally lower 5%.

Fig. 2 Technology
roadmaps without CCS in
China

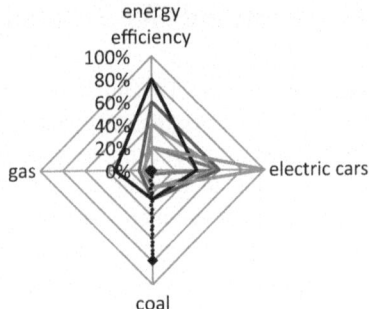

Fig. 3 Technology
roadmaps with CCS in
China

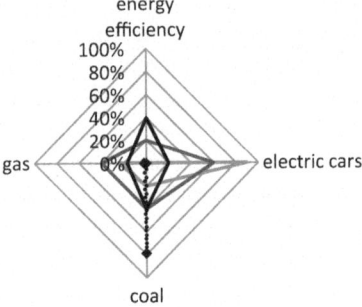

Fig. 2 without CCS and in Fig. 3 with CCS, with the dotted lines representing the 2010 level.

If CCS technology is not applied to the power plants, more efforts should be made in the transport sector and other sectors. For example, if the share of coal is to be reduced from 78.7% in 2010 to 25% in 2050, and the share of gas is to be increased from 1.7% to 33%, then 40% electric vehicles should be employed, and 80% energy efficiency should be improved. Otherwise, if 60% of vehicles are replaced by electric vehicles and the energy efficiency is improved by 20%, the power sector must be almost entirely decarbonized.

However, if all the power plants are equipped with CCS, when 60% of vehicles are replaced by electric vehicles and the energy efficiency is improved by 20%, the reduction of coal can be less than without CCS, from 78.7% to 40%, with the share of gas increased to 20%.

We now propose two roadmaps based on two distinct criteria: one with balanced CO_2 emissions reduction across sectors; the other with least changed energy mix in the power sector.

Fig. 4 Technology roadmaps by balanced technology development in China in 2050 compared to 2010

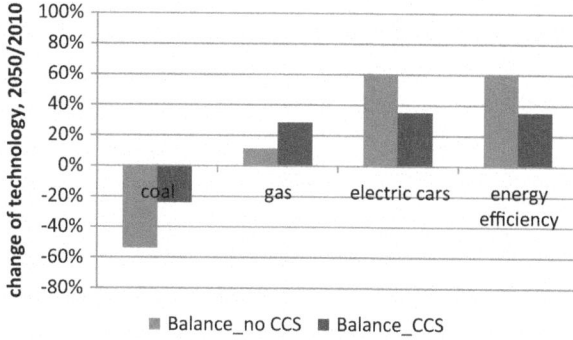

4.1.2 Balanced Technology Development Roadmaps for China

Under the balanced technology development criteria, the technologies across sectors are supposed to be developed homogenously. Figure 4 presents the roadmaps in this criterion with and without CCS for China in 2050. In order to achieve the governmental targets of reducing 90% of CO$_2$ intensity in 2050 with balanced technology development across sectors, electric vehicles in road transport should replace 60% of traditional automobiles. The energy efficiency in the residence and industry sector should be improved by 60%. In the meantime, the share of coal used in the power generation is to be reduced by about 60%, from 78.7% to 25%, thus the share of gas can be increased from 1.7% to 13% in 2050.

However, if CCS technology is implemented in power plants, there will be less direct CO$_2$ emissions reductions in the transport and other sectors. 35% of vehicles will be replaced by electric vehicles, and the energy efficiency in the residence and industry sector should be improved by 35%, nearly less than half of the roadmaps without CCS. In the power sector, coal combustion will be less reduced, from 78.7% to 55%, and the use of gas can be increased to 30% as gas produces less emissions.

4.1.3 Least Changed Energy Mix Roadmaps for China

Now we find out the technology roadmaps by changing the energy mix as little as possible, considering the use of their energy sources as much as possible. Thus more effort will be made in the transport and other sectors.

If CCS is not implemented, the coal in the energy mix in power generation in 2050 can not stay at the same level as in 2010 even with maximum effort of the two other sectors, as shown in the Fig. 5. Actually, if all cars are replaced by electric ones, and energy efficiency is improved by 90% in the residence and industry, the share of coal will have to be reduced by 21.7% (from 78.7% to 57%), with the share of gas remained at 1.7%.

However if all power plants are installed with CCS technology, it is possible that the energy mix stays at the same level in 2050. In transport and other sectors, less

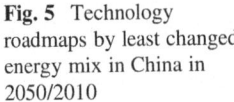

Fig. 5 Technology roadmaps by least changed energy mix in China in 2050/2010

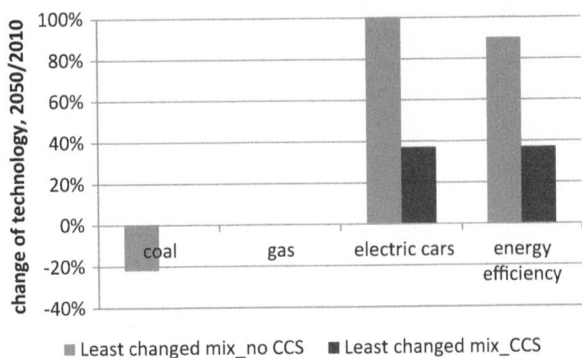

reduction of CO_2 emissions are expected than with CCS. When there is no change in the energy mix in the power sector, 37% of cars have to be replaced by electric cars in the transport sector, and energy efficiency in the residence and industry should be improved by 37%.

4.2 Technology Roadmaps in France

In France, nearly 80% of electricity is now produced by nuclear power. CO_2 emissions from the power sector account for 15% of the total emissions in 2010. The shares of the coal, oil and gas are less than 5% respectively. Thus in the technology roadmaps for France, we focus on the transport and other sectors since the major efforts must be supported by these latter sectors. CCS is not a prior option for France as the CCS is principally installed with the power plants (different from the other two countries).

The transport sector is still the most important sector when it come to contributions to emissions reductions. These technology pathways are presented in the Fig. 6. Energy efficiency in the residence and industry should be improved by at least 40%, and at least 80% of vehicles should be replaced by electric vehicles. For example, if energy efficiency is improved by 80%, 80% of vehicles should be changed to electric vehicles to reach the government target. If all the vehicles are replaced by electric vehicles, energy efficiency is expected to improve by 40%.

As in France the share of coal and gas are very small, the power sector will not contribute much to CO_2 emissions reductions. In this section, we only discuss the balanced technology development roadmaps as the roadmaps with balanced technology development and least energy mix are very similar.

Fig. 6 Technology roadmaps without CCS in France

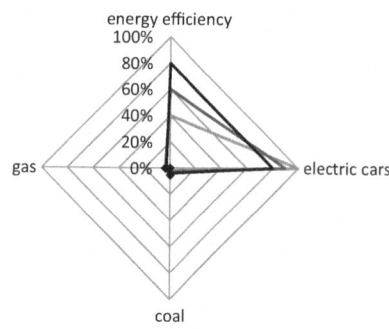

Fig. 7 Technology roadmaps by balanced technology development in France in 2050/2010

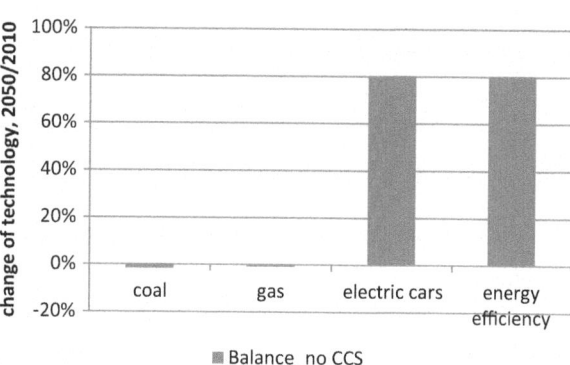

4.2.1 Balanced Technology Development Roadmaps for France

In France, as the share of coal, oil, and gas are already all less than 5%, the power sector can not make much contribution to the reduction of emissions. Most CO_2 emissions will be reduced in the power sector and other sectors. In Fig. 7, we can see that in order to reach the governmental target, 80% of the cars should be replaced by electric cars and the energy efficiency should be be increased by 80%. In these conditions, the share of coal should be reduced by 1.7%, from 5.3% to 3.5%, and the share of gas by 0.9%, from 3.9% to 3%.

4.3 Technology Roadmaps in the US

In the US, 43% of CO_2 emissions come from the power sector in 2010, with the share of coal used in the power sector at 45%, and the share of gas at 23%. The second largest source of emissions was the transport sector, accounting for 30% of total CO_2 emissions in 2010. According to government policy, emissions will be reduced by 82% compared to the level in 2010. Its technology pathways in the policy

Fig. 8 Technology
roadmaps without CCS in
the US

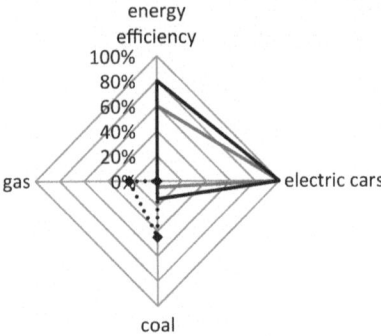

Fig. 9 Technology
roadmaps with CCS in
the US

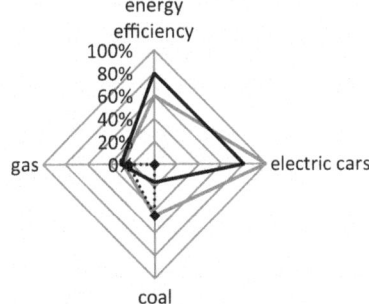

scenario are shown in Fig. 8 without CCS and in Fig. 9 with CCS. In order to achieve
this objective, all the electric vehicles should be replaced by electric vehicles if no
CCS is applied. In this case, either energy efficiency is improved by 60% and the
power sector should be nearly decarbonized, or energy efficiency is to be improved
by 80% and the share of coal is to be reduced at 14%.

If CCS is implemented with all power plants, when energy efficiency is improved
by 60% and all vehicles are replaced by electric vehicles, the share of coal can be
kept at 45% with the share of gas at 28%. When energy efficiency is improved by
80% and 80% of vehicles are changed to electric, the share of coal is to be reduced to
16% with the share of gas at 30%.

4.3.1 Balanced Technology Development Roadmaps for the US

The power and transport sectors are the two most important sectors for CO_2
emissions. They account for 43% and 30% of total emissions in 2010 respectively.
Figure 10 shows the technology roadmaps with balanced technology development
across sectors with and without CCS. If CCS is not considered in the energy
transition in the US, 85% of vehicles should be replaced by electric vehicles, and
the energy efficiency in the residence and industry sector should be improved by

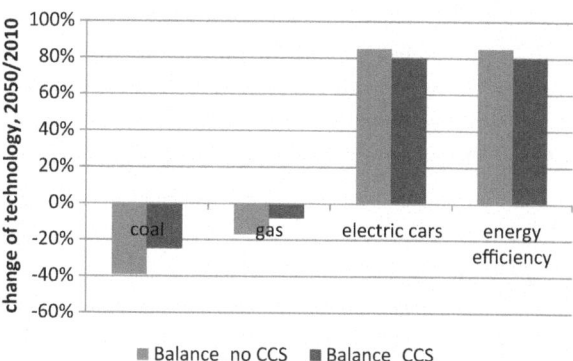

Fig. 10 Technology roadmaps by balanced technology development in the US in 2050/2010

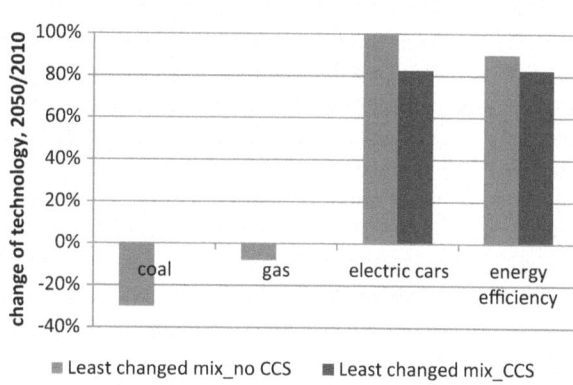

Fig. 11 Technology roadmaps by least changed energy mix in the US in 2050/2010

85%. In the power sector, the share of coal will be reduced by 39%, from 45% to 6%, and the share of gas should be reduced by 17%, from 23% to 6%.

If CCS is implemented in all power plants, the direct reduction of emissions should be slightly less than that without CCS: 80% of cars should be replaced by electric cars, and energy efficiency should be improved by 80%. Meanwhile, more fossil fuels can be used in the power sector than without CCS. The share of coal should be reduced by 25%, from 45% to 20%, and the share of gas should be reduced by 8%, from 23% to 15%.

4.3.2 Least Changed Energy Mix Roadmaps in the US

The technology roadmaps with the least change of energy mix are shown in Fig. 11. When CCS is not implemented, all vehicles should be replaced by electric vehicles and energy efficiency in the residence and industry sector should be improved by 90%. Even with the large reduction in the transport and other sectors, the energy mix in the power sector can not stay at the same level as in 2010. In the power sector, the share of coal will be reduced by 30%, from 45% to 15%, and the share of gas will be reduced by 8%, from 23% to 15%.

If CCS is implemented in all power plants, it is possible that the energy mix in the power sector remains unchanged. Under this condition, 82.5% of cars should be replaced by electric vehicles, and energy efficiency should be improved by 82.5%.

5 Conclusion

In this chapter, a sectoral model has been set up for CO_2 emissions between 2010 and 2050, in China, France and the US, in order to assess the government targets of CO_2 mitigation (UNFCCC 2015). Within the technology roadmaps allowed to accomplish the government targets, electric vehicles were firstly evaluated because of the use of these vehicles will strongly affect electricity production in the power sector. Then the energy mix in each country was assessed based on its energy structure, its electricity output and the choice of the CCS. Finally the improvement of energy efficiency in residence and industry was also calculated. Let us now summarize the main orders of magnitude in energy sectors in these three countries which make it possible to achieve the government's objectives in reducing CO2 emissions.

First, as the most important sector of emissions in France, the transport sector, is expected to contribute two thirds of emission reductions. The low emission vehicles prove indispensable for reaching the French governmental target. According to our results, 80% of vehicles should be changed to electric vehicles in France in 2050. Meanwhile, the energy efficiency in the residences and industries should be improved by 80%. As the power sector contributes little to the CO_2 emission, CCS is not a necessary option for France.

Second, in the US, all sectors should make important contribution for reaching the governmental target. In the transport sector, at least 80% of cars should be changed into electric cars because of the large number of cars in the country, and the energy efficiency in the residences and industries should be improved by at least 60%. In the power sector, the use of coal in 2050 must decrease at most half of that in 2010.

Third for the government target in China, the advancement of technologies is less demanded than in France or in the US. In China, half of the emissions reductions contributions are expected to come from the power sector, as the power sector accouts for half the the emissions in 2010. The power sector and the transport sector should also strongly mitigate emissions in order to achieve the target. For example, if 60% of vehicles can be replaced by electric vehicles, then energy efficiency should be improved by 60%, and coal utilization should be reduced by 60%.

If the energy mix is expected to be remain rather unchanged in China and in the US, CCS should be implemented to all power plants to reach the goals. However this technology is not yet largely applied in the power sector and industry considering of security and high, although the potential of the CCS would prove high in the future.

In this chapter we have focused on the main emission mitigation technologies without considering the cost-effectiveness of these technologies. In future works, the costs of these technologies must be evaluated because they impact the deployment of these new technologies. For instance the cost of batteries of electric vehicles is a very important factor in the technology development and adoption of customers. The next Chapter of the book will focus on this special issue.

Acknowledgment The authors wish to thank J.C. Bocquet and the LGI/CentraleSupélec members for constant supports and especially J. Liu for reviewing our regressions.

Appendix 1: The Framework of the Sectoral Emission Model

The technologies are shown in dotted line in the Fig. 12. We principally analyze the power and transport sectors here, where fuel mix, CCS and electric vehicles are three key factors for CO$_2$ mitigation. Improved energy efficiency in the domestic and industrial sector also are contributors.

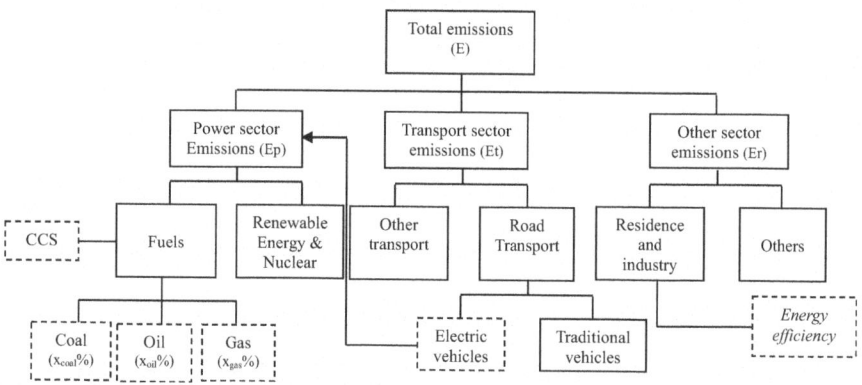

Fig. 12 Schema of the sectoral emission model

Appendix 2: Simulation of Electricity Outputs

SVR is used to provide the underlying function in each country. In our work, the data sets are all normalized from the raw data. We use a sigmoid kernel function for electricity-production prediction. The Polynomial kernel Function is used as the kernel function for electricity output by trial and error. The values of the related hyper-parameters are also turned with a Grid Search. Details regarding the tuning of the parameters and kernel functions can be found in (Liu et al. 2013). The parameters are listed in the Table 3.

The electricity-production simulation results are based on the data of 1971–2010 (IEA 2012). Figures 13, 14 and 15 show the projection of electricity production in the three countries between 1981 and 2050. The X-axis is in years and the Y-axis is electricity output in tWh.

Table 3 Values of the hyper-parameters in electricity output

	C	Degree	ξ	Υ	R^2
China	1	4	$1.0E^{-3}$	10	0.6478
France	1	4	$1.0E^{-3}$	10	0.7161
US	1	4	$1.0E^{-3}$	10	0.7196

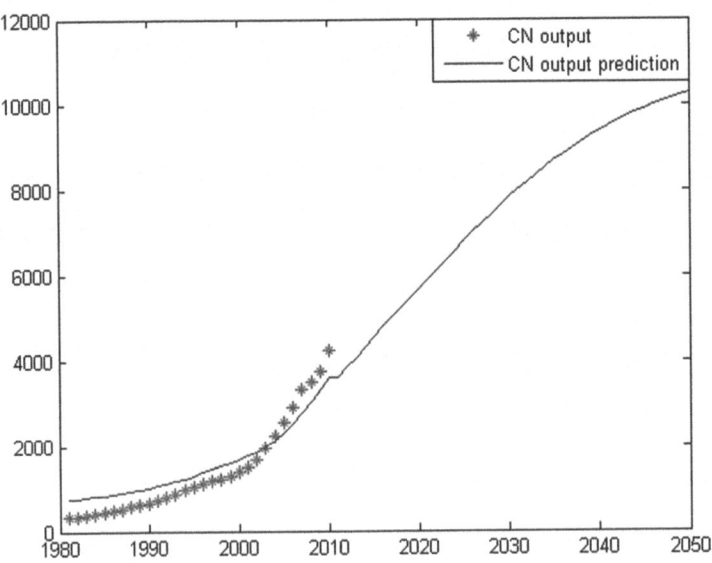

Fig. 13 Electricity production in China 1981–2050

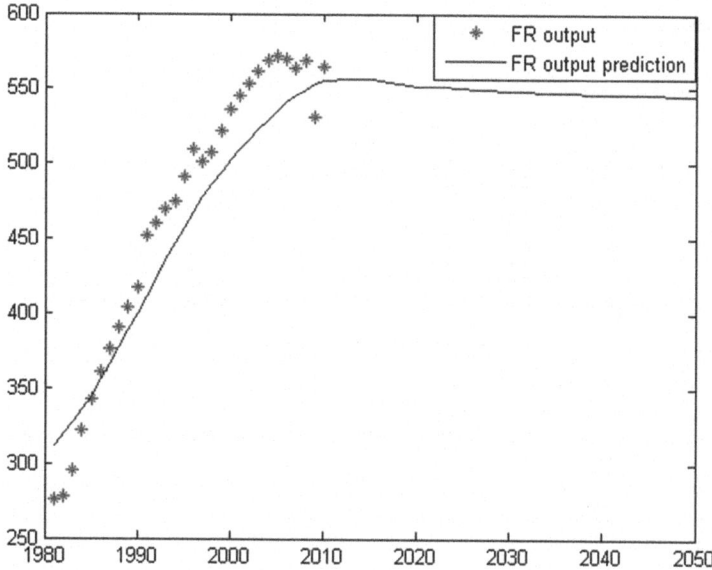

Fig. 14 Electricity production in France 1981–2050

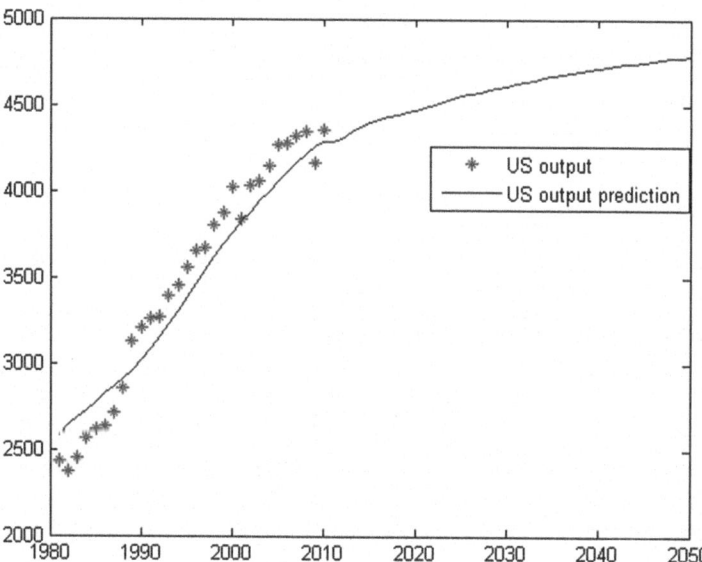

Fig. 15 Electricity production in the US 1981–2050

References

ADEME. (2014). Chiffres clés climat air énergie, 2014.

Akimoto, K. (2016). Evaluations on the emission reduction efforts of Nationally Determined Contributions in cost metrics. Marrakech COP22: Japan Pavilion.

Alazard-Toux, N., Criqui, P., Devezeaux De Lavergne, J.-G., Hache, E., Le Net, E., Lorne, D., Mathy, S., Menanteau, P., Safa, H., Teissier, O., & Topper, B. (2014). Les scénarios de transition énergétique de l'ANCRE. *Revue de l'Energie, 619*, 189–210.

Boulanger, P.-M., & Bréchet, T. (2005). Models for policy-making in sustainable development: The state of the art and perspectives for research. *Ecological Economics, 55*, 337–350.

Cao, L. (2003). Support vector machines experts for time series forecasting. *Neurocomputing, 51*, 321–339.

Chen, W. (2005). The costs of mitigating carbon emissions in China: findings from China MARKAL-MACRO modeling. *Energy Policy, 33*, 885–896.

Cortes, C., & Vapnik, V. (1995). Support-vector networks. *Machine Learning, 20*, 273–297.

ERI. (2009). China's low carbon development path by 2050: Scenario analysisi of energy demand and carbon emissions.

Gao, J. B., Gunn, S. R., Harris, C. J., & Brown, M. (2002). A probabilistic framework for SVM regression and error bar estimation. *Machine Learning, 46*, 71–89.

Hong, W. C. (2010). Application of chaotic ant swarm optimization in electric load forecasting. *Energy Policy, 38*, 5830–5839.

HSBC. (2011, January 4). *The world in 2050: Quantifying the shift in the global economy*. HSBC Global Economics.

IEA. (2008). *Energy techonology prospective 2008: In support of the G8 Plan of Action, Scenario & Strategies to 2050*. OECD.

IEA. (2011). *Technology roadmaps: China wind energy development roadmap to 2050*. OECD/IEA and Energy Research Institute.

IEA. (2012, November 13). *CO2 emission from fuel combustion, highlights*. 2012 Edition.

IEA. (2014, June 1). *Energy technology perspectives 2014: Harnessing electricity's potential*. OECD, Energy Technology Perspectives 2014.

IPCC. (2005). *IPCC special report: Carbon dioxide capture and storage*. Prepared by Working Group III of the Intergovernmental Panel on Climate Change. Cambridge University Press.

IPCC. (2007). *Climate change 2007: Impacts, adaptation and vulnerability*. Contribution of Working Group II to the Fourth Assessment Report of the Intergovernmental Panel on Climate Change. Cambridge University Press.

IPCC. (2013). *Climate change 2013: The physical science basis*. Contribution of Working Group I to the fifth assessment report of the intergovernmental panel on climate change. Cambridge University Press.

Kaya, Y. (1989). *Impact of carbon dioxide emission control on GNP growth: Interpretation of proposed scenarios*. IPCC Response Strategies Working Group Memorandum.

Klaassen, G., & Riahi, K. (2007). Internalizing externalities of electricity generation: An analysis with MESSAGE-MACRO. *Energy Policy, 35*, 815–827.

Liu, J., Seraoui, R., Vitelli, V., & Zio, E. (2013). Nuclear power plant components condition monitoring by probabilistic support vector machine. *Annals of Nuclear Energy, 56*, 23–33.

MEDDE & CDC Climat Recherche. (2015). Les chiffres clés du climat France et Monde.

RITE. (2015). RITE GHG mitigation assessment model DNE21+, system analysis group.

Saveyn, B., Paroussos, L., & Ciscar, J.-C. (2012). Economic analysis of a low carbon path to 2050: A case for China, India and Japan. *Energy Economics, 34*(3), 451–458.

United Nations Framework Convention on Climate Change: UNFCCC. (2015). INDCs as communicated by Parties.

Wang, J., Zhu, W., Zhang, W., & Sun, D. (2009). A trend fixed on firstly and seasonal adjustment model combined with the ε-SVR for short-term forecasting of electricity demand. *Energy Policy, 37*, 4901–4909.

Waxman, H.A., & Markey, E.J. (2009). The American Clean Energy and Security Act of 2009.

Part II
Eco-innovation and New Production Models

Business Model Design: Lessons Learned from Tesla Motors

Yurong Chen and Yannick Perez

Abstract Electric vehicle (EV) industry is still in the introduction stage in product life cycle, and dominant design remains unclear. EV companies, both incumbent from the car industry and new comers, have long taken numerous endeavors to promote EV in the niche market by providing innovative products and business models. While most carmakers still take 'business as usual' approach for developing their EV production and offers, Tesla Motors, an EV entrepreneurial firm, stands out by providing disruptive innovation solutions. We review the business model approach in the literature, then classify the innovation dimensions in the EV ecosystem. We study Tesla Motors in terms of: (1) innovation related to the vehicle, (2) innovation related to the battery (3) innovation concerning the recharging system, and (4) innovation toward the EV ecosystem.

Lessons for incumbent carmakers for their EV business model design: Tesla Motors (1) holds a product strategy entering from high-end market and moving to mass market, with a high level of innovation adaptation and learning by doing; (2) pays considerable attention to reduce range anxiety by high performance super-charger station network and high capacity battery; (3) shows a very high level of integration of information technology into many aspects of the EV business model, such as advanced in-car services and digital distribute channel; (4) shows a new value configuration which involving in high level of vertical integration towards battery and recharging network.

Keywords Business model · Electric vehicle · Tesla motors · Innovation management

Parts of the chapter was made available at: https://www.researchgate.net/publication/277675857_Business_Model_Design_Lessons_Learned_from_Tesla_Motors (Yurong, C., Perez, Y. (2015). Business Model Design: Lessons Learned from Tesla Motors).

Y. Chen · Y. Perez (✉)
Laboratoire Genie Industriel, CentraleSupélec, Université Paris-Saclay, Gif-sur-Yvette, France
e-mail: yurong.chen@centralesupelec.fr; yannick.perez@u-psud.fr

© Springer International Publishing AG, part of Springer Nature 2018
P. da Costa, D. Attias (eds.), *Towards a Sustainable Economy*,
Sustainability and Innovation, https://doi.org/10.1007/978-3-319-79060-2_4

1   Introduction

In the current disruptive period, established business models are under attack from new and incumbent firms with innovative business models. The supply side driven logic of the industrial era that only focus on technology innovation is no longer viable, rather, a successful business model becomes indispensable to convert technology innovation to high firm performance (Baden-Fuller and Haefliger 2013; Chesbrough 2007). Business model innovation does not discover new products or services, instead, it redefines the existing product/service and the way it is provided to the customer. Successful business model innovation can enlarge the existing economic pie, either by attracting new customers or by encouraging existing customers to consume more (Markides 2006). Therefore, business model innovation could set challenges to incumbent firms in matured industries, and also, plays a critical part in the process of commercializing emerging technologies to a new dominant design (Hung and Chu 2006). Business models have the potential to enable the technology advantages which can then be translated into a valuable market offering despite the technology still being immature, and, if proven successful, help gaining a competitive advantage (defensive position) for the firm in the long run (Chesbrough and Rosenbloom 2002). Therefore, business model innovation is congruous with a firm's survival and success for emerging technology as well as industry.

The electric vehicles (EVs, hereafter) industry, or electromobility, has been emerging for near a century, with a series of stops and starts in its development (Donada and Lepoutre 2016; Donada and Perez 2015). The current reintroduction of EV was triggered by high oil prices, climate protection concerns, battery technology and recharging infrastructure development, and the rise of organized car sharing and inter-modality (Dijk et al. 2013). EVs are believed to play an important part in the near future according to policy makers, carmakers and stakeholders (International Energy Agency 2016; MacDougall 2013). Ambitious regional and national goals have stimulated the progress of EV penetration by subsidies for the vehicle and corresponding infrastructure deployment (Dijk et al. 2013). In the year 2016 along, 28 different models of electric vehicle were available in the U.S. market and, among those, 13 are pure battery electric vehicle (BEV, hereafter) models (PluginCars.com 2016). However, the commercialization of EVs has been ineffective thus far, sales of EV are far from satisfactory and lag behind national goals. In 2015, 548,210 EV units (of which BEVs were 60%) were sold globally, which is near double than the sales of 2014, i.e. 317.895 units (EV Sales 2016). While worldwide car sales are expected to reach 742.4 million units in 2015 EV, represented less than 0.07% of the global vehicle market (Statista 2016). Furthermore, the dominant design is still unclear in the EV industry. EV firms are introducing diverse products with diverse business model competing to establish a 'dominant design' (Chen et al. 2016). Accordingly, the EV industry is still in the introductory stage of product life cycle, and struggling to take advantage of economies of scale in small niche markets. EV enterprises, including incumbent and entrepreneurial carmakers, have long undertaken promoting EV in the niche markets by providing innovative business models and overcoming technological shortcomings such as range anxiety. Bohnsack et al. (2014) studied how the path dependencies of incumbent and new entrance firms affected the business models for EVs. And Wang and Kimble (2013) studied the business models of Chinese EVs. Research on how EV companies empirically innovate on business model help us understand how firm solving the complex and radical changing system (Von Pechmann et al. 2015), and bring insights to the industry.

We focus our study on exploring a single case (Yin 2013): Tesla Motors (Tesla, hereafter). Tesla is viewed as a black horse in the auto-industry. Compared with the incumbent auto companies who have decades-experience in making and selling cars, Tesla was a new entrant founded in 2003 by Silicon Valley engineers. Therefore, Tesla has less inert as other incumbent automakers for business model innovation. Tesla is dedicated to the EV-sustainability scenario with innovative products and business models. The product of Tesla, sportive EV Roadster and Model S changed people's idea of the EV and re-initiated the enthusiasm for pure EVs (Urban 2015). Compared to incumbent firms, entrepreneurial firms are generally less constrained and more flexible in pursuing radical technology and business models (Bohnsack et al. 2014; Hill and Rothaermel 2003). While most carmakers still take a 'business as usual' approach towards developing their EV production and offers, Tesla Motors stands out by providing radical innovation solutions (Markides 2006). As a result, we are concerned about the business model design of Tesla and draws several lessons for more incumbent carmakers in their business model design of EV.

This paper starts with presenting the emerging EV industry and business models in the literature, then classifies these innovative dimensions in the EV industry. By combining these two points, a business model innovation framework for EV is developed in Sect. 2. Section 3 is dedicated to reviewing and analyzing the business model innovations of Tesla. Section 4 follows up with the conclusion and recommendations for more classical carmakers.

2 Background and Literature

2.1 Context of Emerging EV Industry

We are currently witnessing the re-introduction of electrical vehicles (EVs) into automobile markets. Unlike the last enthusiasm for EV in 1990s, when the carmakers mainly focused on technological innovations and aimed at providing EV products. In the current EV enthusiasm, the carmakers focuses on many different dimensions, including technology innovations, user relations as a community (e.g. vehicle-to-grid services and car-sharing) and business models innovations (Donada and Lepoutre 2016). This new scenario of EV development is also referred to as electromobility or electromobility 2.0 (Donada and Attias 2015; Donada and Lepoutre 2016). Electromobility remains a nascent industry, where players are currently searching and competing for business models, dominant design, and defining the EV market

(Theyel 2013). Additionally, the network of suppliers, and its players, is in no way stable (Donada and Lepoutre 2016; Fournier et al. 2012).

The scope of the EV industry is much larger than it was in the 1990s: with the connection of the recharging system, EVs are at the intersection between the traditional car making sector and the electricity sector (Chen et al. 2016). The transition into an electric mobility trajectory will lead to fundamental changes in the value chain/ecosystem of the automobile which basically involves components from suppliers, core components and assembly from carmakers, and energy utilities.

First of all, some modules such as the internal combustion engine (ICE) will become less important in the long-term (Huth et al. 2013). While modules such as batteries, charging infrastructure will enter the value chain and play critical roles as a result of high cost and changing peoples' driving behavior (Kley et al. 2011; Weiller and Neely 2014). Secondly, new services enabled by EVs such as energy services or those enlarged by EVs such as car-sharing services and connective services will have numerous influences in the auto value chain (Fournier et al. 2012). At the moment, customers facing services such as energy services and mobility services still await for EV penetration and changes in electricity grid regulation and consumer behavior (Codani et al. 2014a; Weiller and Neely 2014). As a result, the current EV value chain emphasizes on batteries (battery cell manufacturing and battery packing), vehicle (EV design, assembling and sales), and infrastructure enabling grid connection (infrastructure manufacturing and infrastructure network deployment) as is showed in Fig. 1.

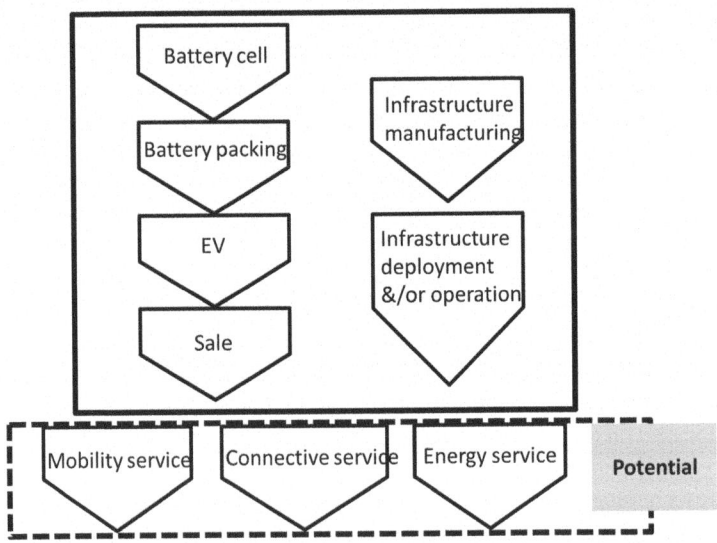

Fig. 1 EV ecosystem, adapted from (Fournier et al. 2012; Weiller and Neely 2014)

2.2 Business Model Innovation in Emerging Industry

The term 'business model' came along with the new challenges and opportunities in the business environment due to new communication technology and computer technology such as the social networks (Osterwalder 2004; Zott et al. 2011). The main goal of a business model is to understand how firm create value and capture value (Chesbrough 2007; Günzel and Holm 2013; Teece 2010; Zott et al. 2011) and they convert payments received on profits (Günzel and Holm 2013; Osterwalder 2004). With business model, we can understand the company's strategy and economic point of view, the statement of market reality, customer expectations, and technological prospects (Baden-Fuller and Haefliger 2013).

The business model is also linked to the company's performance as a result. However, it does not guarantee long-term competitive advantage as other competitors may imitate these practices (Teece 2010). Thus, the creation of a differentiation business models is considered a long term competitive advantage and can set a defensive position for a firm for imitating (Baden-Fuller and Haefliger 2013; Teece 2010). This is also the case since competitors are likely to find it harder to imitate or replicate an entirely new business model than an innovative product or service. With the emergence of a new industry, business model innovation can trigger the commercializing process to find an industrial dominant design and shape the patterns of industrial evolution (Hung and Chu 2006). Therefore, innovation literature treats business model innovation as a cornerstone of transforming technology innovation into a business offering of value (Chesbrough and Rosenbloom 2002; Christensen 1997).

2.3 An Operational Business Model Approach

We applied the business model approach developed by Osterwalder (2004) (known as the business model Canvas (Osterwalder and Pigneur 2010)) and research (e.g. Chesbrough 2010; Günzel and Holm 2013). Osterwalder's mapping of business models, based on extensive literature research, and real-world experience, utilizes nine elements to clarify the processes underlying business models. It contains:

1. Value propositions: defines the promised value of the firm's bundled products or services as well as complementary value-added services. These are packaged and offered by the manufacturer to fulfill customer needs beforehand;
2. Consumer segment: defines the type of customers a company wants to address;
3. Channel: defines how a company delivers the product and services to target customers. It includes direct channels such as through a sales force or over a website, and indirect channels such as reseller and dealer network;
4. Customer relationship: the relationships established with clients;
5. Revenue model: defines what type of payment the customer makes to the supplying shareholder in order to get the product or services.

6. Key partnerships: describes the network of suppliers and partners that make the business model work
7. Capability: are based on a set of resources from the company or its partners to implement the business model
8. Value configuration: defines the potential possibilities to design the product offered with regard to the different shareholders involved in a business model, it has three kinds of configurations which are value chain, value shop and value network. According to the main actors of the car industry, the value configuration is achieved by value chain.
9. Cost structure: describes all costs incurred in operating a business model

This business model mapping illustrates a value creating, delivering and capturing process in a company. While customer segments, channels and customer relationship are obviously value delivering processes (Günzel and Holm 2013), channels can also contribute to value creation—online shopping could bring convenience as a value for example to customers by shipping-to-destination services. Value proposition is critical for value creation, and partnerships, capability and value configuration are indispensable tools to make value creation happen (Osterwalder and Pigneur 2010). Value configuration is also related to value capturing, since it determines what value added activities a firm will perform and is highly linked to the cost structure of firm. Revenue model and cost structure are of great interest in such a business model, especially for executers and investors, as it is connected to profits profile and has a central place in the value capturing process (Günzel and Holm 2013).

3 Methodology

3.1 Case of Choice and Data Collection

We chose the case of Tesla for two reasons. The first is that in the field of electric vehicles, Tesla has already been recognized as a strong agent of change. Its flagship vehicle, Tesla Model S, was the world's best-selling plug-in car in 2015 (EV Sales 2016), and its share price has surged since 2013 (NASDAQ.com, 2016), indicating high customer satisfaction and investor expectations. Second, Tesla is a entrepreneurial company and established at the Silicon Valley, a cluster for innovations. Therefore, Tesla has less inert than incumbent automakers for business model innovation and could take more radical trajectory for innovation (Hill and Rothaermel 2003). Its business model stands out and attracts attention from business researchers (e.g. Bohnsack et al. 2014; Weiller et al. 2013). Third, Tesla is very open and transparent of their activities and strategies by posting on the official website and blogs, while the incumbent carmaker are very strict to keep the information in secret. As a most popular EV makers, Tesla is very well-documented by the media, which facilitates the collection of rich and often real-time data.

Our a single case (Yin 2013) is based primarily on secondary qualitative data. We used secondary sources, which are abundantly available for the chosen cases as previously explained. We collected and analyzed data from the official website and annual reports of Tesla (e.g. Tesla Motors 2013, 2016); books such as *Owning Model S: The Definitive Guide to Buying and Owning the Tesla Model S* (detailed information on the products of Tesla); and *Elon Musk: Tesla, SpaceX, and the Quest for a Fantastic Future* (information on the vision of Elon Musk, the CEO of Tesla); blogs for Tesla (where Elon Musk posts regularly); and reports of industry associations and magazines such as *Automotive News*, *Ward's AutoWorld*, *Autoweek*, and *Electric Cars Report*. The data was collected for the period from June 2011 (when Tesla went public) to June 2016. In addition to these sources, we also analyzed academic case studies on Tesla (e.g. Donada and Lepoutre 2016).

3.2 Business Model Innovation Frame in EV Ecosystem

We apply the business model frame adapted from Osterwalder (2004) to analyze business model innovation in EV. The EV industry involves new modules and components as a result of battery-based electric mobility concepts, such as recharging infrastructure and related services. In the EV ecosystem, early studies have identified three dimensions for business models: vehicle together with battery; the infrastructure system; the system services which integrated electric vehicles into the energy system (Kley et al. 2011). However, regarding the current business and research of EV, electricity system services (e.g. Vehicle to Grid, Vehicle to Home) is in the very early stage of the life-cycle (Theyel 2013), where only researches and prototypes take place (Codani et al. 2014b; Weiller and Neely 2014). In this vein, we adapted the key dimensions of EV business model innovation into the following:

1. Innovation towards the vehicle;
2. Innovation towards the battery;
3. Innovation towards the infrastructure system;

We add another dimension which is the EV ecosystem in our analysis, more precisely, value configuration in the ecosystem.

4. Innovation towards the ecosystem.

We apply the business model mapping of Osterwalder (2004) to analyze the innovations in Tesla. Among the nine elements in the mapping, we select five (value proposition, value configuration, channel, consumer segment and revenue model).

4 Findings

4.1 Innovation Towards Vehicles:

Tesla motor has thus far released four vehicle models into market: a two doors sport car Tesla Roadster (2008–2012), a sedan Tesla Model S (2012-), a crossover Tesla Model X (2015-) and a family car Tesla Model 3 (2016-). The vehicles received high attention from the public and the media, because they address the high end customer segment, which are new for EVs, and its innovative multi-channel for distribution.

4.1.1 Value Proposition

Musk (2006) declaimed that "Critical to making that [EV becoming mainstream] happen is an electric car without compromises, which is why the Tesla Roadster is designed to beat a gasoline sports car like a Porsche or Ferrari in a head to head showdown". Tesla's first car, the Roadster, released in 2008, changed people's imagination of EV, which was small-size and low-speed. Roadster looks like a fancy sport car, using the body of Lotus Elites. At the same time, it also offers fast-speed and powerful acceleration as well as high performance in the range for one charge, which is an important parameter for EV. Range anxiety is one of the serious problems facing EV makers and EV users- EV users are afraid they cannot reach their destination and run out of battery. It can reach 100 km/h within 3.7 s acceleration and a standard range of 393 km with a one-time charge. An EV usually has an autonomy of less than 100 km, and has an image of small-size low-speed vehicle.

Following the success of the Roadster, Tesla released Model S in 2012, with purposed vehicle design for a premium family car. The intersection between aesthetics and performance attracted popularity from both customers and investors. The Model S range has a range from 335 km to 426 km, depending on the version, and with an acceleration speed as fast as 2.8 s (duel motor version), which is much faster than most luxury sport cars. Model S won many awards and honours such as "most stylish car in Switzerland", "best inventions of the year", and "Automobile of the Year" (DeMorro 2015).

Model X was released on the market on September 2015. It uses falcon wing doors for access to the second and third row seats, which gives a stylish appearance. The range and acceleration speed is similar to Model S.

Half a year later, Tesla unveiled its 4th Model 3, which is a compact sedan targeting lower segments compared to Model S and X. Yet, it choose a stylish design and "aesthetics will not be sacrificed" (Hull 2016). It offers range of 346 km and 0–100 km/h acceleration less than 6 s. As of 7 April 2016, 1 week after the unveiling, company officials said they had taken 325,000 Model 3 reservations, more than triple the number of Model S cars Tesla had sold by the end of 2015.

Tesla emphasises connective technology and self-driving technology. Tesla innovatively increased the connectivity between users and the environment (e.g. recharging navigation stations, charging control and autopilot) enabled by IT based hardware and software applications. It innovatively offers data network in the car with telecommunication partners, and connects the car with the maintenance centre, infotainment centre and so on.

4.1.2 Customer Segment

Tesla entered the market of EV by targeting the high-end niche market, by offering a luxury specific-purpose vehicle such as Roadster. Model S targets luxury the multi-purpose car market as a result sales are considerably larger than the Roadster. Furthermore, it continues to offer an SUV version luxury multi-purpose car, followed by a more economical multi-purpose car. It corresponds to the strategic goal of creating an affordable mass market EV. The customer segments of battery and recharging systems need to match the customer segment of vehicle.

The customer segment is vastly different to other carmakers which usually enters from a multi-purpose economy or specific-purpose market as the ownership cost for EV is high (Bohnsack et al. 2014).

4.1.3 Distribution Channel

As a newcomer to the car industry, Tesla Motors changed the conventional dealership network for vehicle distribution. It created a new multi-channel model for purchasing vehicles, which involved online stores and apple-like retail outlets. The online stores offer potential customers the chance to purchase the car directly online. The retail outlets are usually located in dense traffic, enhanced with technology which has high integration of IT in order to better present Tesla vehicle and its company culture. Tesla applies vertical integration on sales, which means the price of vehicles is unnegotiable.

4.1.4 Revenue Model

Tesla applied an ownership-as-usual model for the revenue. They the sell the car to individuals, and as a result, the customers possess the ownership of the car (other than a mobility service without car ownership). Tesla also sells powertrains and battery packs to other carmakers as a supplier to their EVs. For example, Tesla and Daimler have an agreement over battery packs and chargers for Smart Fortwo from

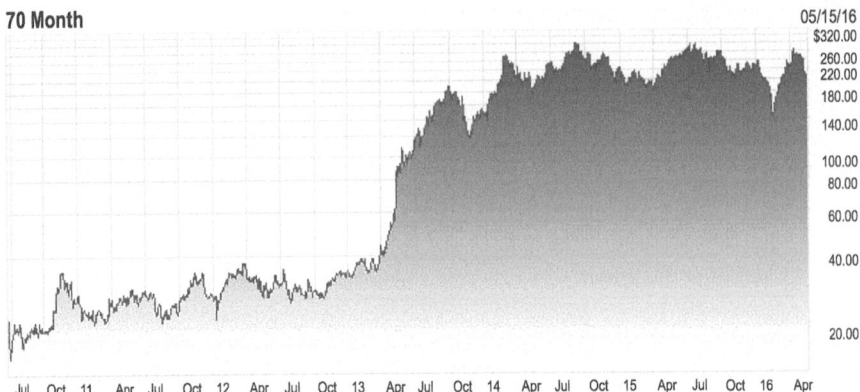

Fig. 2 Stock market of Tesla. Source: http://www.nasdaq.com/symbol/tsla, accessed May 15th 2016

2008 to 2013, and develops powertrain systems for Toyota RAV4 from 2010 to 2014.

Other types of revenue include government loans and investment such as in stock markets. In 2010, Tesla received US government loan for development and production of Model S (which has been paid back at 2013). Besides government loan, Daimler spent $50 million in 2009 for a 10% stake of Tesla, and Toyota bought $50 million worth of stock when Tesla went public in July 2010. The outstanding performance on stock markets brings further capital (Fig. 2).

4.2 Innovation Towards the Battery

In 2013, an electrical powertrain with a 10 kWh battery pack takes around 57% of the value-add in all components in an EV. And the average rate of added value for conventional powertrain is 26% (Huth et al. 2013). The choice of battery will largely decide the range anxiety and the cost that customers will have. Tesla applied an ambitious plan on battery strategy, with expecting movements on battery factory and enter the stationary battery market. It is attractive for its high range, and innovative battery pack technology.

4.2.1 Value Proposition

Starting with Roadster, Tesla innovatively chose battery packs with large capacities as a solution to range anxiety issues. The Roadster was equipped with a 53 kWh battery and has autonomy of 393 km. Such capacities significantly exceed those of any other commercially available electric vehicle at the same time, for example, in 2009, the BMW MINI E chose a battery pack of 35 kWh with a range of 160 km, and iMiEV in 2010 offered a battery pack of 16 kWh and a range of 100 km. This outstanding feature continues in Model S and Model X. In 2016, the new versions of Model S has a battery pack options of 70 kWh and 90 kWh that provide a range of 335 or 426 km, respectively.

Tesla motor has a good knowledge of battery packs and management system. It has innovatively equipped Roaster with thousands of laptop Lithium-ion cells and assembles them into a performance and cost optimized battery pack. During the delivery of Tesla Model S, it developed a closer relationship with its battery cell supplier Panasonic, on both battery technology and the scale of production.

The connectivity service can link users to battery packs to some extent. Tesla users can have some control on the battery system. For example, users can control the temperature of the battery system before entering the car when the environmental temperature is too low.

4.2.2 Distribution Channel and Revenue Model

The battery is generally sold to customers along with the vehicle, with possibility for extra purchase when the old one is at the end of life and need to be replaced. As previously mentioned, Tesla also sells its innovative battery pack to other companies.

4.3 Innovation Towards Infrastructure System

Another ambitious plan of Tesla Motors is the expansion of the supercharger network. It is famous for its high performance in charging ability, well-established networks and free to Tesla user strategy.

4.3.1 Value Proposition

In alignment with the large battery capacity adapted by Tesla, the supercharger station offers fast charging in order to satisfy the charging needs of customers. It can deliver direct current up to 120 kW and capable of charging to 80% of an 85 kWh Tesla Model S within 40 min. Besides the premium function of the supercharger station, Tesla is undertaking an ambitious expansion plan to establish a network of superchargers along well-traveled highways and in congested city centers. Until May 2015, there were 2400 superchargers in 400 stations worldwide. One year later, there are 3708 superchargers in 624 stations in May 2016.

Tesla also has a pilot project for a battery swap program, it was launched in several regions to meet the charging needs of customers and reduce range anxiety. All of the superchargers are connected to Tesla, and users can access it via the screen in the car. Tesla users can find the nearest supercharging station and control the charging when connected.

4.3.2 Distribution Channel

The public network is solely deployed by Tesla Motors. This is mainly due to the different charging technology and standard adapted by the companies, and the different cables that are designed and adapted.

4.3.3 Revenue Model

Tesla users benefit from free entrance to the supercharger stations network. However, Tesla needs to bear all the cost including installment, maintenance and network reinforcement if needed. The rent for the place is shared by a supercharger partner program with local partners.

4.4 *Innovation Towards Ecosystem*

In the conventional car industry, the value chain consists in the pyramid relationship between the carmaker and suppliers, in which suppliers provide the different parts or modules such as the gearbox and auxiliary battery to carmakers, while the main role of carmakers is assembling the parts and designing core competents such as motor design as well as the vehicle body; on the other hand, energy utility will fill the car with fuel during the car's lifetime as showed in Fig. 3a. A classic carmaker in-house production share is around 25% for the total vehicle (Huth et al. 2013).

In the EV industry, most carmakers who are engaging in the EV market choose to follow their old routine of value configuration: they tend to use their existing production infrastructure, capabilities, as well as supplier network (Chen et al. 2016). In this type of value chain, carmakers treat battery as a module for outsourcing, it could be because of the limitation on technological knowledge or transaction cost concern. BMW i3 and Renault Zoe are examples as showed in Fig. 3e,f respectively. A better choice could be the carmaker and battery supplier form an joint venture company, as it is the case for Nissan leaf (Fig. 3d). On the other hand, as for the recharging network deployment, most carmakers wait for the action from the recharging operation company or other stakeholders such as national or local governments. Renault and BMW followed this strategy, and their EVs are able to access to the recharging network deployment by chargepoint and chargemaster in USA and UK. Furthermore, BMW has started to invest in the fast recharging infrastructure

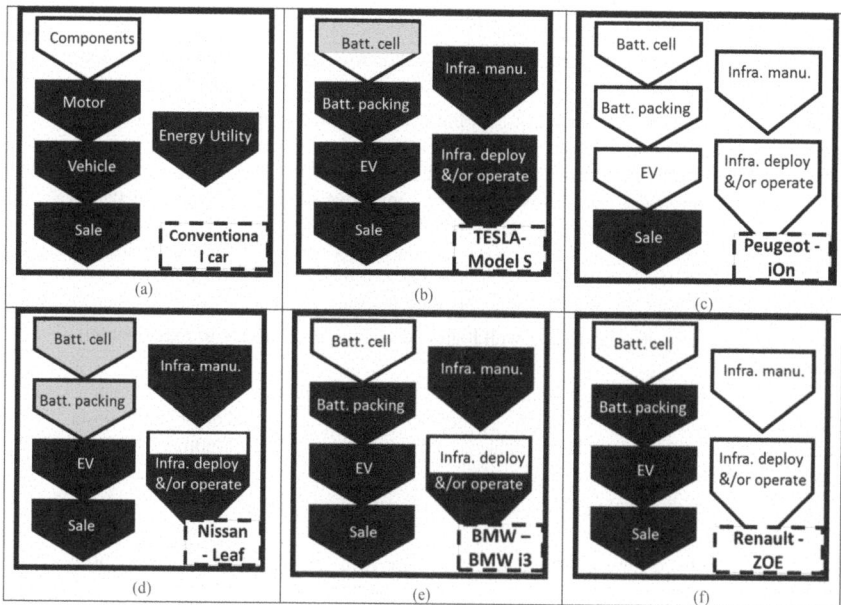

Fig. 3 Value configurations of Tesla Motors and other carmakers (black- outsource from supplier/ other utility; grey- joint venture; white- Vertical integration by carmaker)

network with partners as of end of 2014 (Fig. 3e,f). Nissan started developing quick charging networks in 2012, earlier and more aggressively than BMW, but still by partnership with utility providers (Fig. 3d). At the same time, companies which are less engaged in the EV market thus far, who wish to keep EV in their product portfolio could choose to be less integrated in their value chain, and purchase the EV from another carmaker. As Citroën C-Zero and Peugeot iOn from PSA are examples for this type of value configuration, it purchases the i-MiEVs from Mitsubishi, and resale it in europe under the brand Citroën and Peugeot. As a result, PSA only occupies the sale position in the value chain of EV (Fig. 3c).

In contrast, Tesla shows a very high different value configuration compare to other carmakers, from high level of out-sourcing to high level of in-house making. During the delivery time of the Tesla Roadster, most components are outsourcing to the suppliers, including battery cell, vehicle design and manufacturing. It is mainly due to that the company is in the initial stage, and in lack of knowledge and capacities for vehicle production and fast repond to the market. However, the packing and assembling of the battery cells and the energy management are conducted by Tesla. When the commercial delivery of Tesla Model S began, Tesla motors began to show a high level of vertical integration along its value chain: body design, battery packing, recharging system as well as recent move towards batter cell manufacture as the Gigafactory with Panasonic (Fig. 3b).

Therefore, a map for the business model innovation of Tesla Motors is summarized in (Table 1).

Table 1 Business model of Tesla Motors from value-related perspective

	Innovation towards vehicle	Innovation towards battery	Innovation towards infrastructure system
Value proposition	High performance regarding to range and vehicle performance; innovative connective services and intelligent services	Innovative management of battery packs enables high capacity and low cost; connective service enables interaction with users; new products towards stationary battery market	High performance recharging station with highly developed recharging station network; connective service enable interact with user;
Customer segments	Innovatively starting with high-end market; and moving to mass market		
Distribution channel	Innovative multi-channel model, involving high integration of IT; vertical integration on selling	Together with vehicle, replace possible	Public network deployed by tesla motors only
Value configuration	Innovatively possess high level of vertical integration		
Revenue model	Ownership; government loan	Purchase with vehicle or separate purchase when update	Free to tesla users
	Selling powertrain and battery pack to other EV maker		
	Market share		

5   Conclusion

This paper discusses the business model innovation of Tesla Motors regarding vehicle, battery, infrastructure systems and their corresponding value configurations. Following the analysis, we arrived on a systematic view of how Tesla innovates in the business model.

A top-down and flexible product strategy: Tesla Motors holds a product strategy entering from high-end market and moving to mass market customer segments. It started with offering performance sport EV which ignited the market enthusiasm, followed by providing the premium family EV and aiming to create affordable mass market for EV. At the same time, as an entrepreneurial firm, it has a high level of innovation adaptation and flexibility in learning by doing. More classical carmakers should also be more flexible in pursuing radical business models, especially when the dominant technology design in EV industry are unclear.

A huge endeavor on range anxiety reduction: Tesla Motor holds plan to solve the range anxiety problem along with EV. It pays a considerable attention to both large capacity battery packs and high performance supercharger stations. One of the most important long term strategies of Tesla Motors is the high performance supercharger station and its aggressive expansion around the main intercity highways in US and Europe. Furthermore, the strategy choice of battery range is much higher than the choice of other carmakers. All these aspects contribute to reducing the range anxiety of Tesla users and enable high performance in the value proposition of business model. As range anxiety comes with the attributes of EV and become the most critical concern for the customer, carmakers should also take certain actions to reduce the range anxiety with certain cost.

An integration of information technology: Tesla shows a high level of integration of information technology into the EV business model. In the value proposition, Tesla innovatively increased the connectivity between users and the environment such as charging stations and infotainment services. Tesla benefits from the attackers' advantage in the connectivity of car (Christensen and Rosenbloom 1995). A high share of information technology is involved in both online and retail outlet distribution channels for Tesla. The connective service will increase the add-on-value of vehicle or after sell services, carmakers should take action on integrating information technology for both the vehicle value proposition and distribution channel.

A new value configuration with more integration: Tesla Motor holds a new value configuration which involves a high level of vertical integration towards battery and recharging network. The integration strategy will reduce coordinate costs between carmakers and their suppliers, and reduce risks caused by lack of supporting infrastructure. However, it also involves high investment and risk coming from the uncertainty of the EV industry.

Acknowledgements Yurong Chen benefits from the support of the Chair "PSA Peugeot Citroen Automobile: Hybrid technologies and Economy of Electromobility", so-called Armand Peugeot Chair led by CentraleSupélec, and ESSEC Business School and sponsored by PEUGEOT CITROEN Automobile. She would like to express her gratitude towards all partner institutions within the program as well as the Armand Peugeot Chair. An early version of this paper has been presented in Gerpisa annual conference 2015.

References

Baden-Fuller, C., & Haefliger, S. (2013). Business models and technological innovation. *Long Range Planning, 46*(6), 419–426. https://doi.org/10.1016/j.lrp.2013.08.023.

Bohnsack, R., Pinkse, J., & Kolk, A. (2014). Business models for sustainable technologies: Exploring business model evolution in the case of electric vehicles. *Research Policy, 43*(2), 284–300. https://doi.org/10.1016/j.respol.2013.10.014.

Chen, Y., Chowdhury, S., Donada, C., & Perez, Y. (2016). *Mirroring hypothesis and integrality in the electric vehicle industry: Evidence from Tesla Motors.* Administrative Sciences Association of Canada Conference, Montréal.

Chesbrough, H. (2007). Business model innovation. It's not just about technology anymore. *Strategy & Leadership, 35*(6), 12–17. https://doi.org/10.1108/10878570710833714.

Chesbrough, H. (2010). Business model innovation: Opportunities and barriers. *Long Range Planning, 43*(2–3), 354–363. https://doi.org/10.1016/j.lrp.2009.07.010.

Chesbrough, H., & Rosenbloom, R. S. (2002). The role of the business model in capturing value from innovation: Evidence from Xerox Corporation's technology spin-off companies. *Industrial and Corporate Change, 11*(3), 529–555. https://doi.org/10.1093/icc/11.3.529.

Christensen, C. M. (1997). The innovator's dilemma. *Harvard Business School Press*, 1–14. https://doi.org/10.1016/j.jbusres.2010.12.002.

Christensen, C. M., & Rosenbloom, R. S. (1995). Explaining the attacker's advantage: Technological paradigms, organizational dynamics, and the value network. *Research Policy, 24*(2), 233–257. https://doi.org/10.1016/0048-7333(93)00764-K.

Codani, P., Petit, M., & Perez, Y. (2014a). *Missing money for EVs: Economics impacts of TSO market designs.* Available at SSRN https://ssrn.com/abstract=2525290

Codani, P., Petit, M., & Perez, Y. (2014b). *Diversity of transmission system operators for Grid Integrated Vehicles.* In 11th International Conference on the European Energy Market (EEM14) (pp. 1–5). IEEE. https://doi.org/10.1109/EEM.2014.6861209.

DeMorro, C. (2015). *How many awards has Tesla won? This infographic tells us.* Accessed May 17, 2016, from http://cleantechnica.com/2015/02/18/many-awards-tesla-won-infographic-tells-us/

Dijk, M., Orsato, R. J., & Kemp, R. (2013). The emergence of an electric mobility trajectory. *Energy Policy, 52,* 135–145. https://doi.org/10.1016/j.enpol.2012.04.024.

Donada, C., & Attias, D. (2015). Food for thought: Which organisation and ecosystem governance to boost radical innovation in the electromobility 2.0 industry? *International Journal of Automotive Technology and Management, 15*(2), 105–125. https://doi.org/10.1504/IJATM.2015.068545.

Donada, C., & Lepoutre, J. (2016). How can startups create the conditions for a dominant position in the nascent industry of electromobility 2.0? *International Journal of Automotive Technology and Management, 16*(1), 11–29.

Donada, C., & Perez, Y. (2015). Editorial – Electromobility at crossroads. *International Journal of Automotive Technology and Management, 15*(2), 97–104.

EV Sales. (2016). *World top 20 2015 special edition.* Accessed May 11, 2016, from http://evsales.blogspot.fr/2016/01/world-top-20-december-2015-special

Fournier, G., Hinderer, H., Schmid, D., Seign, R., & Baumann, M. (2012). The new mobility paradigm. Transformation of value chain and business models. *Enterprise and Work Innovation Studies, 8*(IET), 9–40.

Günzel, F., & Holm, A. B. (2013). One size does not fit all – Understanding the front-end and back-end of business model innovation. *International Journal of Innovation Management, 17*(1), 1340002–1340034. https://doi.org/10.1142/S1363919613400021.

Hill, C. W. L., & Rothaermel, F. T. (2003). The performance of incumbent firms in the face of radical technological innovation. *The Academy of Management Review, 28*(2), 257–274. https://doi.org/10.2307/30040712.

Hull, D. (2016). *Tesla says it received more than 325,000 model 3 reservations.* Accessed May 17, 2016, from http://www.bloomberg.com/news/articles/2016-04-07/tesla-says-model-3-pre-orders-surge-to-325-000-in-firstweek

Hung, S. C., & Chu, Y. Y. (2006). Stimulating new industries from emerging technologies: Challenges for the public sector. *Technovation, 26,* 104–110. https://doi.org/10.1016/j.technovation.2004.07.018.

Huth, C., Wittek, K., & Spengler, T. S. (2013). OEM strategies for vertical integration in the battery value chain. *International Journal of Automotive Technology and Management, 13*(1), 75. https://doi.org/10.1504/IJATM.2013.052780.

International Energy Agency. (2016). *Global EV outlook 2016: Beyond one million electric cars.* https://www.iea.org/publications/freepublications/.../Global_EV_Outlook_2016.pdf

Kley, F., Lerch, C., & Dallinger, D. (2011). New business models for electric cars—A holistic approach. *Energy Policy, 39*(6), 3392–3403.

MacDougall, W. (2013). *Electromobility in Germany: Vision 2020 and beyond.* Berlin: Germany Trade & Invest.

Markides, C. (2006). Disruptive innovation: In need of better theory. *Journal of Product Innovation Management, 23,* 19–25. https://doi.org/10.1111/j.1540-5885.2005.00177.x.

Musk, E. (2006). *The secret Tesla Motors master plan: Blog – Tesla Motors.* Accessed January 9, 2017, from https://www.teslamotors.com/blog/secret-tesla-motors-master-plan-just-between-you-and-me

Osterwalder, A. (2004). *The business model ontology: A proposition in a design science approach.* Lausanne: HEC Lausanne.

Osterwalder, A., & Pigneur, Y. (2010). *Business model generation: A handbook for visionaries, game changers, and challengers.* https://doi.org/10.1523/JNEUROSCI.0307-10.2010.

Statista. (2016). *Global car sales by manufacturer 2015.* Accessed May 11, 2016, from http://www.statista.com/statistics/271608/global-vehicle-sales-of-automobile-manufacturers

Teece, D. J. (2010). Business models, business strategy and innovation. *Long Range Planning, 43* (2–3), 172–194. https://doi.org/10.1016/j.lrp.2009.07.003.

Tesla Motors. (2013). *2012 annual report of Tesla Motors Inc.* California.

Tesla Motors. (2016). *2015 annual report of Tesla Motors Inc.* California.

Theyel, G. (2013). A framework for understanding emerging industries. In *EAEPE conference.* Paris.

Urban, T. (2015). *How Tesla will change the world.* Accessed May 12, 2016, from http://www.businessinsider.com.au/how-tesla-will-change-the-world-2015-6/

Von Pechmann, F., Midler, C., Maniak, R., & Charue-Duboc, F. (2015). Managing systemic and disruptive innovation: Lessons from the Renault zero emission initiative. *Industrial and Corporate Change, 24*(3), 677–695.

Wang, H., & Kimble, C. (2013). Business model innovation in the chinese electric vehicle industry. In G. Calabrese (Ed.), *The greening of the automotive industry* (pp. 240–253). London: Palgrave Macmillan.

Weiller, C., & Neely, A. (2014). Using electric vehicles for energy services: Industry perspectives. *Energy, 77,* 194–200. https://doi.org/10.1016/j.energy.2014.06.066.

Weiller, C., Shang, A., Neely, A., & Shi, Y. (2013). Competing and co-existing business models for EV: Lessons from international case studies. In *Electric vehicle symposium and exhibition* (pp. 1–12). https://doi.org/10.1109/EVS.2013.6914776.

Yin, R. (2013). *Case study research design and methods* (5th ed.). Thousand Oaks, CA: SAGE.

Zott, C., Amit, R., & Massa, L. (2011). The business model: Recent developments and future research. *Journal of Management, 37*(4), 1019–1042. https://doi.org/10.1177/0149206311406265.

Availability of Mineral Resources and Impact for Electric Vehicle Recycling in Europe

Hakim Idjis and Danielle Attias

Abstract Lithium-ion battery technology is a key component of vehicle electrification and its end-of-life recovery is an important factor in lifting barriers towards increased Electromobility, such as battery cost, environmental impact, mandatory recycling rates of more than 50% battery weight (European Union) and, finally, the availability of constituent elements such as lithium and cobalt. This chapter focuses on the availability of constituent materials, in order to assess the potential for critical shortages due to a scaling up of Electromobility. To account for the complexity and long-term horizon of our study, we combine the use of System Dynamics with the Stanford Research Institute Matrix for scenario planning. We find that for lithium-ion battery needs, only cobalt is likely to see its reserves depleted. Other materials such as nickel, manganese, copper, graphite and iron are at risk of depletion due to developments unrelated to Electromobility. In all cases, we show that recycling significantly reduces the consumption of materials for lithium-ion batteries.

Keywords Lithium-ion · Battery · Electric vehicle · Recycling · Criticality

1 Introduction

In this twenty-first century, our planet is facing unprecedented challenges for its preservation; this is especially true of our energy model which must be redesigned to meet growing demand, establish more democratic access worldwide, while minimizing the environmental impact of its production and use (WEC 2013). The transportation sector as a whole, road transport in particular, has a strong impact

H. Idjis
PSA-Citroën, Laboratoire Genie Industriel, CentraleSupélec, Université Paris-Saclay, Gif-sur-Yvette, France
e-mail: hakim.idjis@centralesupelec.fr

D. Attias (✉)
Laboratoire Genie Industriel, CentraleSupelec, Université Paris-Saclay, Gif-sur-Yvette, France
e-mail: danielle.attias@centralesupelec.fr

© Springer International Publishing AG, part of Springer Nature 2018
P. da Costa, D. Attias (eds.), *Towards a Sustainable Economy*,
Sustainability and Innovation, https://doi.org/10.1007/978-3-319-79060-2_5

on our energy model due to its dependence on oil and its contribution to greenhouse gas (EEA 2015; European Commission 2014a; IEA 2012). Making it Greener is therefore a priority.

Measures have been taken at the EU and global levels to reduce these emissions. European commitments call for a transport sector emission reduction of 60% in 2050 compared to emission levels in 1990 (European Commission 2011). European Regulation No. 443/2009 was introduced as part of the climate package. It defined emissions thresholds for new light vehicles up to 130 g of C02 per km by 2015 and 95 g per km in 2020 (European Commission 2009) and was completed in 2014 by the European Regulation No 333/2014 (European Commission 2014b). Today, the 2015 goal is attainable by various car manufacturers (EEA 2015). The 2020 goal remains however, a challenge that requires radical innovation.

Fuel cells and other promising technologies being less mature in terms of technology and infrastructure, the industry has tended to turn towards electric vehicles (hybrid, plug-in hybrid and battery-electric) in order to comply with this regulation and thus reduce European energy dependence and emissions. According to Idjis and Da Costa (2017): "These electric vehicles (EV) mainly use lithium-ion batteries (LIB)", which "give them greater autonomy. However, they crystallize some of the barriers preventing widespread use, such as the cost of the battery, its impact on the overall life cycle assessment of the EV and the availability of constituent materials". It is the latter barrier that we are investigating in this chapter.

EV engines and batteries consume rare soil and strategic materials. Concerns are regularly expressed on dependence, potential depletion and environmental consequences of mining some raw materials such as lithium, cobalt or graphite. According to the USGS (U.S. Geological Survey), 85% of the world's cobalt and lithium production comes from only seven countries (USGS 2010). Among the latter, there are countries in Latin America and sub-Saharan Africa, with unstable political regimes. Thus, the first concern is dependence on producer countries.

The second issue concerns the physical availability of resources, or what we call geological criticality, although for lithium the problem of its availability seems to be solved. Many studies show that there are enough reserves (economically exploitable resources) for the most optimistic EV scenarios (Grosjean et al. 2012; Gruber et al. 2011).

Finally, the third concern is on the ability of producing countries to meet demand in the years to come, given that it takes 5–10 years for a new mine to become exploitable (Miedema and Moll 2013; Novinsky et al. 2014).

Through all these elements of context, we conclude that the battery is the central technological element accompanying vehicle electrification, and that its valorization at the end-of-life is an important lever to lift the brakes on the deployment of Electromobility. In this work, we position ourselves on the battery's end-of-life perimeter, where we will be interested in the effect of recycling on the raw materials consumption and criticality.

Section 2 now introduces the technical scope of battery and recovery technologies: these concepts are necessary for understanding the subsequent analysis. Our approach and results will be detailed in Sects. 3 and 4 respectively. Section 5 concludes this chapter.

2 Battery and Recovery Technologies

2.1 Battery

Several battery technologies are available for electrifying vehicles such as: nickel cadmium (Ni-Cd), nickel metal hydride (Ni-MH) and lithium-ion (LIB). This latter is more adapted and used for automotive applications, given its performances (energy density, voltage of cells, lifespan and memory effect). A LIB is made up of two principal parts: the electrical/support part (battery management system, connecting cables and cooling system), and the electrochemical one (set of modules, which are composed primarily of cells) (see Fig. 1).

Therefore the basic component of a battery is the cell, which gives it its name. A cell consists of an electrolyte, a separator, a cathode (positive electrode) and an anode (negative electrode). Each one of these electrodes is composed of a conductor and an active material. For the anode, the active material is usually graphite. For the cathode, the positive active material is a combination of **lithium and a metal oxide**, which varies from one technology to another. The properties of the LIB are defined by the latter (lifespan, safety, capacity, and cost). In this study, we selected two families of key technologies:

- Nickel-Manganese-Cobalt based batteries (NMC) with the average composition $Li(Ni_{1/3}Mn_{1/3}Co_{1/3})O_2$ (ADEME 2013; Hoyer et al. 2014);
- Iron phosphate based batteries (LFP) with the composition $LiFePO_4$.

In terms of materials contained in a LIB, Fig. 2 shows the average proportions in a pack (Idjis and Da Costa 2017):

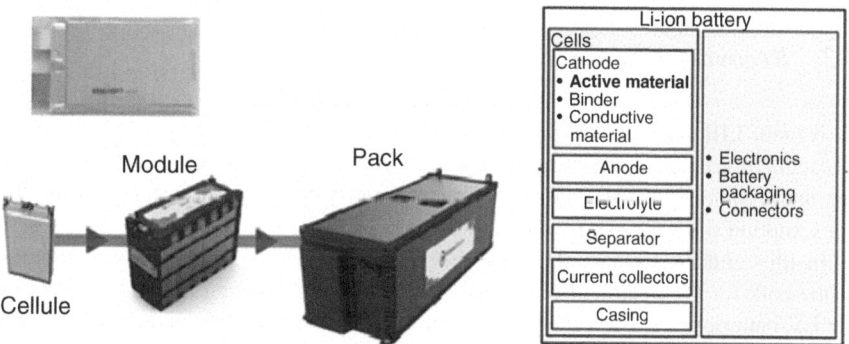

Fig. 1 Composition and decomposition of a battery. (Swart et al. 2014; Väyrynen and Salminen 2012)

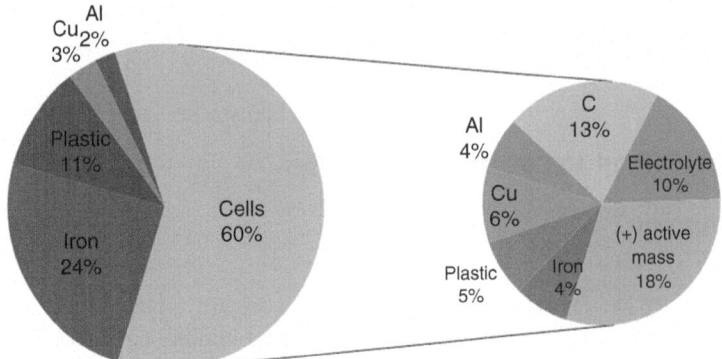

Fig. 2 Analytical decomposition of an EV pack (Idjis and Da Costa 2017)

- Steel and iron: casing of the battery cells, bolts;
- Aluminum and copper: electrode conductors, cables, electronic boards;
- Plastics: casing of the battery, cables, separator;
- Graphite: negative active material of the anode;
- Lithium: electrolyte and positive active material of the cathode;
- Cobalt, Nickel, Manganese, iron, phosphorus: positive active material of the cathode;
- And solvents.

Depending on the level of vehicle electrification, the LIB capacity defines its mass. For hybrid vehicles "HEV", the capacity is between 1 kWh and 2 kWh, which result in a mass of about 30 kg. For plug-in hybrid ones "PHEV", the capacity is between 5 kWh and 15 kWh, which result in an average mass of 150 kg. Last of all, for battery-electric vehicles "BEV", with a capacity over 15 kWh, the average mass is about 250 kg.

2.2 Recovery Options

Two main LIB recovery options need to be considered: recycling (as required by EU regulation) and repurposing for reuse in 2nd life applications. In this work, we are interested in the first one in order to quantify the effect of recycling on the consumption of the constituent materials.

In Idjis and Da Costa (2017), the EU directive 2006/66/EC sets the regulatory framework for the treatment of batteries and accumulators at end-of-life. It imposes for EV batteries: (1) The establishment of a collection and a recycling system; and (2) The requirement to recycle at least 50% of the battery weight. Therefore, the recycling objective is the achievement of these regulatory targets, while recovering the value contained in the LIB materials.

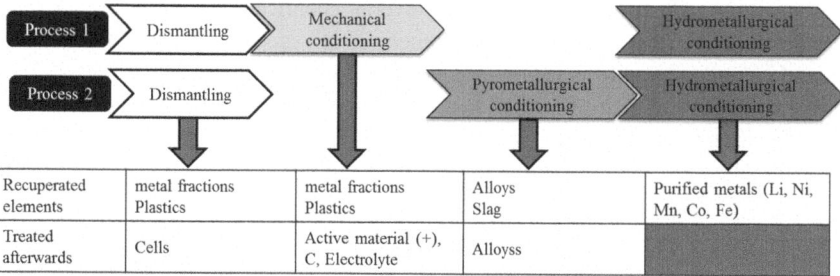

Fig. 3 Considered recycling processes

These materials are found at different levels in the battery (electrode, cell, pack), in varying proportions (remember Figs. 1 and 2), as well as with a different contribution to the recovered value. Operations of extraction, separation and purification are required consequently. Kwade (2010) identifies four possible basic processing operations: Dismantling; Mechanical conditioning; Pyrometallurgical conditioning; and hydrometallurgical conditioning. These operations could be combined in several ways to form five recycling alternatives. The scientific and industrial state of the art by Idjis (2015) considers two recycling processes as shown in Fig. 3. Each process is a succession of three operations, the recovered materials at each operation are described in the bottom. We denote Process 1 by P1 and Process 2 by P2.

Idjis and Da Costa (2017) notice that the recovery of materials contained in the positive active material (which are difficult to access) requires an elaborate recycling process, therefore a high cost. This is why materials such as lithium are not recycled today.

In the future, with the development of EVs, the level of criticality for any material will certainly justify the economic benefit of its recycling, which will reduce this initial criticality. We have highlighted above the so-called System Dynamics methodology (Sterman 2000) i.e. a feedback loop. The concept of feedback loops can be explained using the analogy of vicious or virtuous circles, wherein an influence spreads among several factors and returns back to the factor that initially generated it.

Due to the presence of these feedback loops, we have used System Dynamics Modeling in our approach, which is explained in the next section.

3 Approach

System Dynamics (SD) is a suitable methodology for the analysis of large-scale complex systems wherein heterogeneous factors interact, stemming from systems thinking theory. "The objective is to analyze, understand and predict the behavior of this system over time by analyzing its changing factors" (Sterman 2000). For our study and Idjis and Da Costa (2017), this means identifying the factors that create the dynamics of minerals consumption, modeling of internal laws of behavior between these factors and their time simulations in several scenarios. This was done in a much broader study by Idjis (2015), in which other objectives such as recovery profitability and compliance with recycling targets were investigated (Fig. 4).

As shown in Fig. 4, a SD model is a set of factors related by links of causalities. The Fig. 5 shows a simplified diagram of the SD model developed in (Idjis 2015), regarding the mineral consumption for LIBs in Europe.

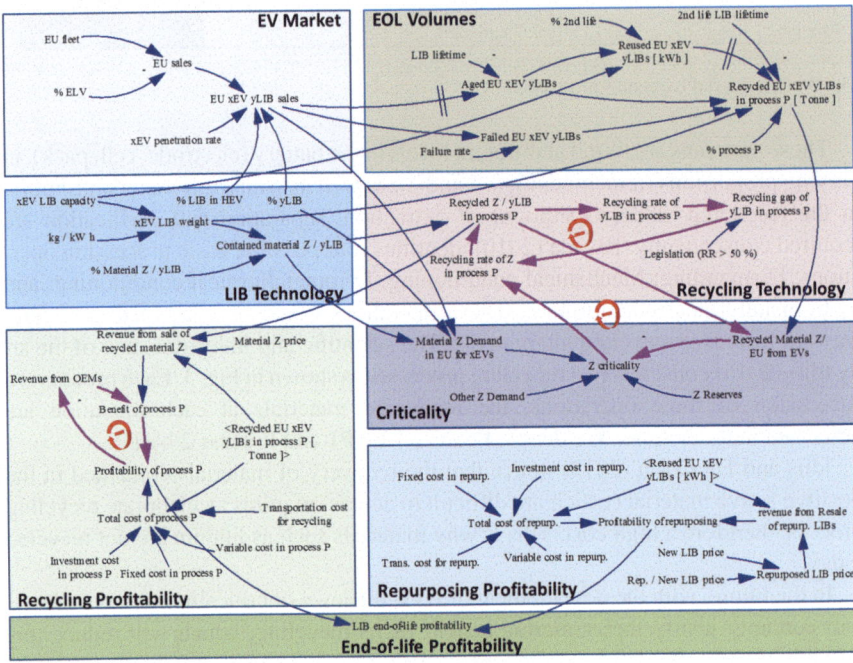

Fig. 4 Overview of the LIB recovery network SD model (Idjis 2015)

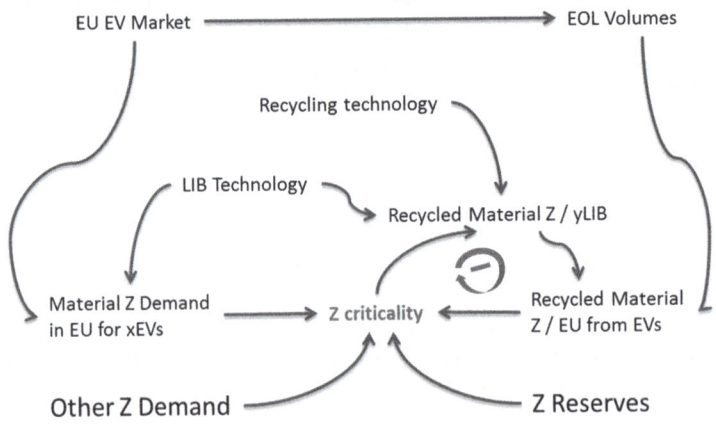

Fig. 5 Simplified SD diagram regarding the dynamic of minerals consumption for LIBs in Europe

Table 1 SRI matrix of scenarios

		80%	0%	% 2nd life
		80%	20%	% NMC
12.5–30 (kWh)	ETP_2DS	S1	S2	
10–24 (kWh)	Mean 2DS—4DS	S3	S4	
LIB capacity PHEV—EV	Penetration rate EV			

The Fig. 5 is as follows: the criticality of any material 'Z' is dictated by its reserves, its consumption in other markets, its consumption for EVs and the recycled amount of this material. The consumption for EVs is calculated based on the EVs market and LIB technologies developments, while the recycled amount is induced by end-of-life volumes and recycling technologies. Finally, we find the feedback loop (red arrow) explained above (criticality—increasing the recycled amount—decreasing criticality).

Being in a prospective study, we have combined the use of SD with scenarios. The choice of the Stanford Research Institute (SRI) matrix is justified given its suitability with the our SD model characteristics: complexity, heterogeneous factors and emergence (Acosta and Idjis 2014; Idjis and Da Costa 2017).

The SRI matrix is a crossing of two dimensions of factors that dictate primarily the dynamic evolution of the SD model. The scenarios introduced in Table 1 are derived from (Idjis 2015).

For example in the S1 scenario, 80% of end-of-life automotive LIBs undergo 2nd life reuse before recycling and the main technology is based on nickel, manganese and cobalt (NMC). For Idjis and Da Costa (2017), the electric vehicles sales volume is from the IEA's 2DS energy scenario (IEA 2012). This latter is based on proactive environmental policies to contain global warming to 2° (2DS) in 2050, unlike the 4° scenario (4DS). These vehicles have a capacity of 30 kWh and 12.5 kWh for EVs and PHEVs respectively. We notice that the most minerals consuming scenarios (pessimistic) are S1 for minerals: Li, Ni, Co, Mn, Cu, Al, C, and S2 for minerals: Fe, P.

The model is simulated from 2010 (first sales of electric vehicles with LIBs) until 2050. This is consistent with the literature on EV sales and minerals consumption (IEA 2012; Miedema and Moll 2013; Pasaoglu et al. 2012). In this timeframe, the LIBs will be the reference technology to be recovered at least until 2040 and beyond with post LIB batteries. As a reminder, we are considering here a European geographical area.

4   Results and Discussions

We begin with a preliminary analysis of geological criticality before the development of Electromobility. To do so, we calculated the number of years remaining for the mining of each material at its 2010 fixed level (Table 2).

Table 2 Number of years remaining for reserves consumption

	Li	Ni	Mn	Co	Cu	Al	C	Fe	P
Reserves (millions t)	13.5	81	570	7.3	700	28,000	110	87,000	67,000
Prod_2010 (millions t)	0.028	1.59	13.9	0.09	15.9	40.8	0.925	1140	181
Nb years (prod_2010)	482	51	41	81	44	686	119	76	370

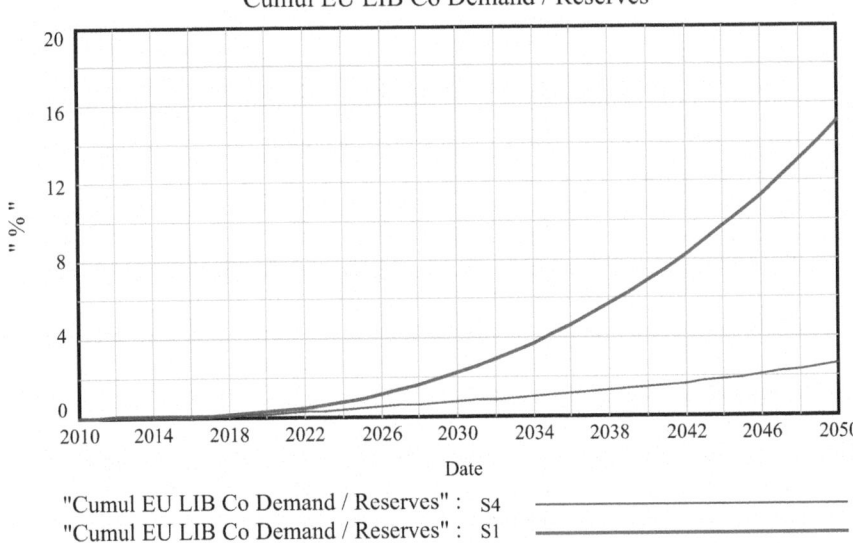

Fig. 6 Cobalt reserves consumption for EU LIBs in S1/S4 scenarios

We note that there are materials for which there is less than a hundred years of exploitation, even before the marketing of electrified vehicles based on LIBs. This is the case of Ni, Mn, Co, Cu and Fe, which present a potential geological criticality before the development of Electromobility. To determine whether this geological criticality is effective and possibly induced by Electromobility, it is necessary to integrate, using the SD model, future demand for LIBs, recycling and future demand in other markets.

Initially, we analyze the effect of LIBs in Europe, without recycling. To do this, we consider the extreme cases. The pessimistic scenarios (consumers of material)/ optimistic (not consumers of materials) are represented by S1/S4 (for materials: Li, Ni, Co, Mn, Cu, Al, C) and S2/S3 (for Fe, P). For example, the Fig. 6 illustrates the result obtained in the DS model on the consumption of cobalt reserves for EU LIBs. The rest of the results are summarized in Fig. 7.

These columns show the ratio between the cumulative demand (2010–2050) of a material for the LIBs needs in Europe and its current reserves. Except for cobalt in S1, no other material presents a risk of geological criticality due to the development of EVs (less than 4%).

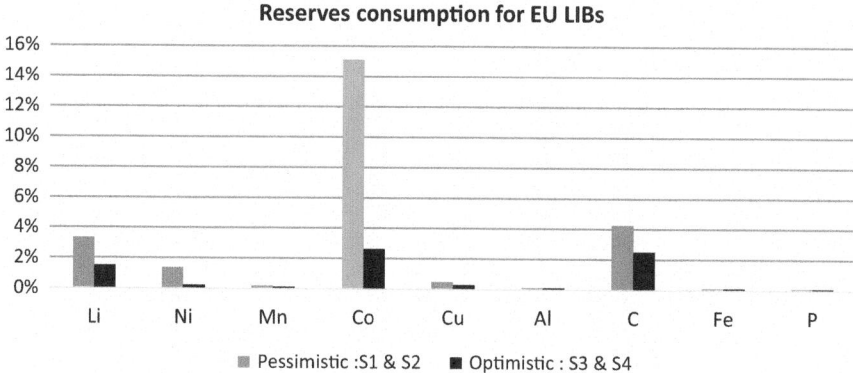

Fig. 7 Consumption of material reserves for electromobility needs in Europe

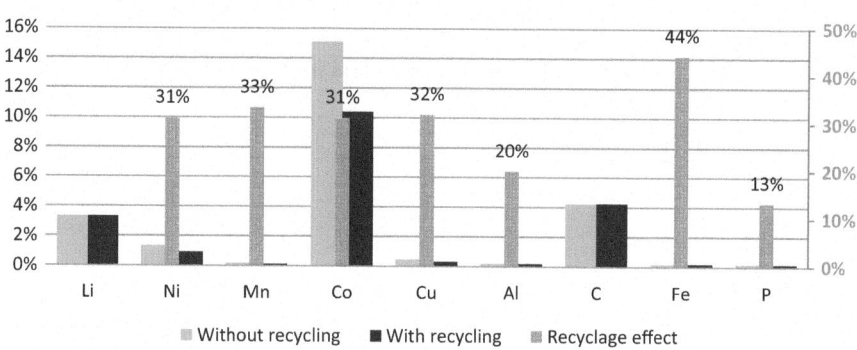

Fig. 8 Recycling effect on reserves consumption for EU LIBs

The Fig. 7 does not consider the recycling of LIBs in Europe, which has the effect of reducing the consumption of reserves. To analyze this effect, we considered the pessimistic case (S1 and S2), it was in this latter that we detected a potential geological criticality of cobalt, as shown in the Fig. 8.

We note that recycling significantly reduces the consumption of some materials (Ni, Mn, Cu, Al, Fe) for LIBs in Europe, although these do not present a risk of geological criticality, due to EVs development. In the case of Cobalt, even with recycling, 10% of the reserves will have been consumed by 2050. To sum up the demand for cobalt for the LIBs in the rest of the world and the demand for the other industrial sectors, one can expect a higher consumption of reserves. This is the purpose of the next section.

At present, we will combine the effect of demand for LIBs with the rest of the demands, including LIBs outside Europe. To do this, we consider only S1 and S2 for the demand from Electromobility. For the other applications, we consider three situations, stagnation (+ 0%/year), moderate increase (+ 2%/year) and strong increase (+ 5%/year). Recycling is included in all cases. The stagnation in demand

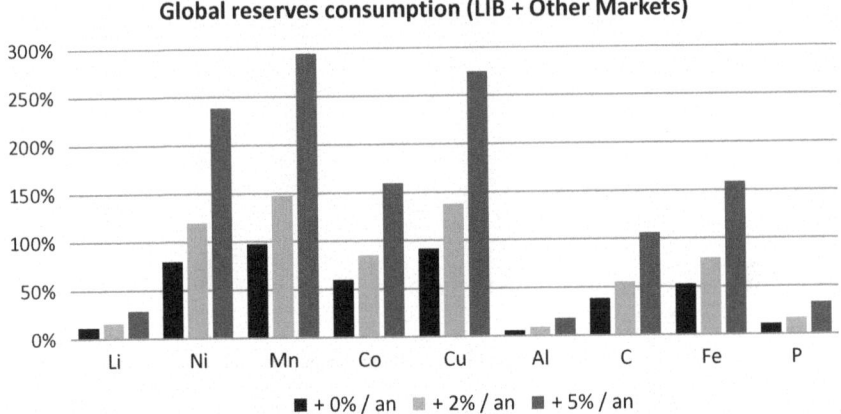

Fig. 9 Global reserves consumption

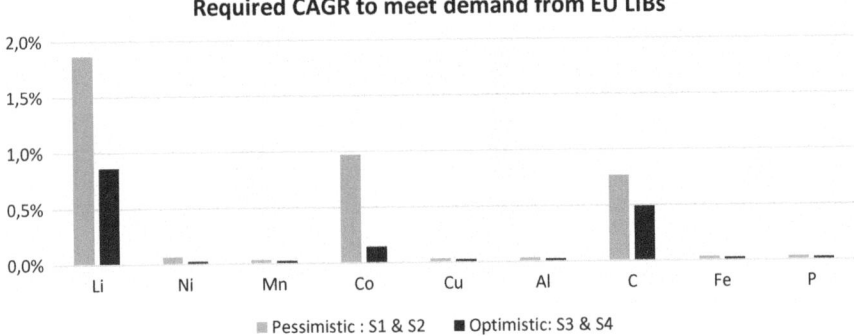

Fig. 10 Necessary production levels needed for electromobility in Europe

(+ 0%/year) is a rather optimistic situation, considering the history of the last 5 years (USGS 2015, 2014, 2013, 2012, 2011, 2010).

By analyzing the three demands profiles, we conclude that materials (Li, Al, and P) do not present a risk of geological criticality. The remaining materials (Ni, Mn, Co, Cu, C, Fe) present this risk and require mitigation strategies. Cobalt is the only material contributing to this risk through the development of Electromobility (Figs. 8 and 9). Mitigation strategies are therefore to be developed outside the automotive sector (increase in the recycling rate, substitution, exploration of new reserves and such).

Before concluding on the geological criticality of materials, we present a final analysis concerning the ability of production to meet future demand, especially in the event of a sudden increase induced by the deployment of EVs. We are interested in this because of the slow process, of the order of 5–10 years for the opening of a new mine (Miedema and Moll 2013; Novinsky et al. 2014). The Fig. 10 shows the required *Compound annual growth rate* (CAGR) in relation to 2010 production to meet the needs of LIBs in Europe, considering recycling.

The demands for materials (Ni, Mn, Cu, Al, Fe and P) for EU LIBs represent a negligible proportion of their production in 2010, even in the pessimistic case. The required annual increase in production to meet EU LIB requirements is less than 0.07% per year. Consequently, it is not the development of LIBs which will lead to a risk of deficit in the production of these materials.

Lithium, cobalt and graphite are the only materials requiring efforts to open new mines. For example, lithium production should be increased by almost 2%/year (1%/year) for LIBs in Europe, in the pessimistic (optimistic) case. By adding the demand for LIBs in the rest of the world and the demand from other industrial sectors, we can expect a necessary increase of up to 10%/year.

The increase in the production of these materials must therefore be anticipated. It will be necessary to ensure that the required production levels in 2020 and 2025 are currently under development. However, we do not have information on the prospective capabilities of mining companies to conduct such an audit.

5 Conclusion

In this chapter, we investigated the effect of the Electromobility market development in Europe on the consumption of materials, especially on the perimeter of the lithium-ion battery, which is the central technological element accompanying the electrification of vehicles. We have also analyzed the effect of recycling to reduce the consumption of these materials and their criticality.

We conclude that for the purposes of the LIBs, only the use of cobalt is likely to exhaust a large part of the reserves. Other materials such as nickel, manganese, copper, graphite and iron present a risk of depletion due to developments beyond the scope of Electromobility. Therefore, in-depth analyzes considering other sectors of use (future demand, substitutability, etc.) are necessary. In all cases, we have shown that recycling can significantly reduce material consumption for LIBs.

Regarding lithium, we have shown, contrary to what is expected of the public, that it does not present a risk of exhaustion even in the most optimistic scenarios of EVs deployment. However, its supply may be disrupted by other risk factors, such as geographic concentration of deposits and associated geopolitical risks, which calls for a multi-criteria approach of criticality estimation.

References

Acosta, C., & Idjis, H. (2014). *State of the art of scenario planning: Proposal of a classification of scenario building existing methods according to use* (Mémoire thématique). Ecole Centrale Paris.
ADEME. (2013). *Élaboration selon les principes des ACV des bilans énergétiques, des émissions de gaz à effet de serre et des autres impacts environnementaux.* Induits par l'ensemble des filières de véhicules électriques et de véhicules thermiques, VP de segment B (citadine

polyvalente) et VUL à l'horizon 2012 et 2020. Agence de l'Environnement et de la Maîtrise de l'Energie.

EEA. (2015). *Global search on data, maps and indicators – European environment agency.* Accessed April 14, 2015, from http://www.eea.europa.eu/data-and-maps/find#c1=Graph& c1=Map&b_start=0&c6=transport

European Commission. (2009). Regulation (EC) No 443/2009 of 23 April 2009 setting emission performance standards for new passenger cars as part of the Community's integrated approach to reduce CO2 emissions from light-duty vehicles.

European Commission. (2011). *White paper on transport: Towards a competitive and resource efficient transport system.*

European Commission. (2014a). *Statistical pocketbook 2014 – Transport.* European Commission.

European Commission. (2014b). Regulation (EU) No 333/2014 of the European Parliament and of the Council of 11 March 2014 amending Regulation (EC) No 443/2009 to define the modalities for reaching the 2020 target to reduce CO2 emissions from new passenger cars.

Grosjean, C., Miranda, P. H., Perrin, M., & Poggi, P. (2012). Assessment of world lithium resources and consequences of their geographic distribution on the expected development of the electric vehicle industry. *Renewable and Sustainable Energy Reviews, 16*, 1735–1744.

Gruber, P. W., Medina, P. A., Keoleian, G. A., Kesler, S. E., Everson, M. P., & Wallington, T. J. (2011). Global lithium availability. *Journal of Industrial Ecology, 15*, 760–775.

Hoyer, C., Kieckhäfer, K., & Spengler, T. S. (2014). Technology and capacity planning for the recycling of lithium-ion electric vehicle batteries in Germany. *Journal of Business Economics, 85*, 505–544.

Idjis, H. (2015). *La filière de valorisation des batteries de véhicules électriques en fin de vie: Contribution à la modélisation d'un système organisationnel complexe en émergence* (Phd thesis). Université Paris-Saclay, Français.

Idjis, H., & Da Costa, P. (2017). Is electric vehicles battery recovery a source of cost or profit? In D. Attias (Ed.), *The automobile revolution* (pp. 117–134). Cham: Springer. https://doi.org/10. 1007/978-3-319-45838-0_8.

IEA. (2012). *Energy technology perspectives 2012: Pathways to a clean energy system.* Paris: International Energy Agency.

Kwade, A. (2010). *On the way to an "intelligent" recycling of traction batteries. Presented at the 7th Braunschweiger symposium on hybrid.* Braunschweigh: Electric Vehicles and Energy Management.

Miedema, J. H., & Moll, H. C. (2013). Lithium availability in the EU27 for battery-driven vehicles: The impact of recycling and substitution on the confrontation between supply and demand until 2050. *Resources Policy, 38*, 204–211.

Novinsky, P., Glöser, S., Kühn, A., & Walz, R. (2014). *Modeling the feedback of battery raw material shortages on the technological development of lithium-ion-batteries and the diffusion of alternative automotive drives.* In 32nd International Conference of the System Dynamics Society, Delft, Netherlands.

Pasaoglu, G., Honselaar, M., & Thiel, C. (2012). Potential vehicle fleet CO_2 reductions and cost implications for various vehicle technology deployment scenarios in Europe. *Energy Policy, 40*, 404–421.

Sterman, J. (2000). *Business dynamics: Systems thinking and modeling for a complex world.* New York: Irwin/McGraw-Hill.

Swart, P., Dewulf, J., & Biernaux, A. (2014). Resource demand for the production of different cathode materials for lithium ion batteries. *Journal of Cleaner Production, 84*, 391–399.

USGS. (2010). *Mineral Commodity Summaries 2010.* U.S. Geological Survey.

USGS. (2011). *Mineral Commodity Summaries 2011.* U.S. Geological Survey.

USGS. (2012). *Mineral Commodity Summaries 2012.* U.S. Geological Survey.

USGS. (2013). *Mineral Commodity Summaries 2013.* U.S. Geological Survey.

USGS. (2014). *Mineral Commodity Summaries 2014.* U.S. Geological Survey.

USGS. (2015). *Mineral Commodity Summaries 2015.* U.S. Geological Survey.

Väyrynen, A., & Salminen, J. (2012). Lithium ion battery production. *The Journal of Chemical Thermodynamics, Thermodynamics of Sustainable Processes, 46*, 80–85.

WEC. (2013). Time to get real – The case for sustainable energy investment. *World Energy Trilemma 2013.* World Energy Council.

Comparing Sustainable Performance of Industrial System Alternatives in Operating Conditions

Yann Leroy and François Cluzel

Abstract Life Cycle Assessment (LCA) assesses the environmental performance of products through their entire life cycle. LCA-based decision-making usually focuses on environmental impact, omitting other considerations, such as customer expectations and economic aspects. Although the environmental performances of industrial systems are highly dependent on operating conditions (e.g. local context, accessibility of resources), LCA usually integrates generic data. The aim of this chapter is to provide an integrated framework to identify the solutions best suited to a specific context, considering environmental, economic and commercial aspects.

First, the environmental performances of competing products are compared using LCA. A scenario-based approach is then applied based on the most influential parameters identified in the environmental assessment. Costs are then incorporated into a set of exploitation scenarios.

Second, several customer profiles are generated with respect to their economic and environmental strategies. Products are then ranked according to these profiles. The final step is to identify the most suitable solution for a given context-client couple.

The framework is applied to three burners for forge furnaces. Results demonstrate that client profiles and operating contexts (namely client expectations, location and resource availability and costs) affect technological choices.

Keywords Sustainable industrial system · Life-cycle assessment · Customer profiles · Exploitation scenarios · Operating local contexts

Parts of this chapter were published in Open Access in 2015 on https://hal.archives-ouvertes.fr/hal-01144384/document (Leroy, Y., Cluzel, F., Lamé, G. (2015) Comparing Sustainable Performance of Industrial System Alternatives by Integrating Environment, Costs, Clients and Exploitation Context, HAL w.p.).

Y. Leroy (✉) · F. Cluzel
Laboratoire Genie Industriel, CentraleSupélec, Université Paris-Saclay, Gif-sur-Yvette, France
e-mail: yann.leroy@centralesupelec.fr; francois.cluzel@centralesupelec.fr

1 Introduction

In designing and defining the most appropriate technological solutions for industrial systems, sustainable issues continue to gain interest. Large energy-consuming industrial systems are of particular concern in this landscape of solutions as they may be implemented worldwide in vastly different operating contexts, with potential economic, environmental and social impacts.

Nevertheless, environmental performance is sensitive to the external factors, i.e. geographical location, accessibility of resources, and regulations (Bertoluci et al. 2014). Significant uncertainties may exist in the life cycle of an industrial system, limiting the ability to obtain accurate Life Cycle Assessment (LCA) results (Cluzel et al. 2014; Reap et al. 2008a, 2008b). Indeed, systems are often complex, featuring a high number of subsystems and parts, relatively long lifespans (30–40 years) and the occasional upgrading. In addition, sustainable decision-making cannot disconnect environmental performance from customer expectations and economic analysis (Norris 2001).

In this chapter, we implement an original, context-based approach combining environmental assessment and customer expectations through LCA, an economic evaluation, which is the most widely recognized and complete approach. Usually limited to environmental or economic evaluation, this integrated approach aims at identifying the best industrial solution for a specific context. This framework is applied to three alternative forge furnace burners.

Economic and environmental impacts may vary from one industrial system's geographical site to another (Smith and Mago 2013). Technology-invested decision-making based on sustainable considerations is thus a hard task. We have already proposed contributions to improve the reliability of the environmental evaluation of such systems using LCA (Cluzel et al. 2014); however only the environmental aspects were considered.

Life Cycle-based decision-making involves combining environmental, economic and customer aspects. In this context, a sustainable framework is proposed to deal with this kind of technological decision-making. The aforementioned technological performance may be highly sensitive to the local operational context since resource availability is of considerable interest.

The proposed framework is illustrated on Fig. 1. First, we define a set of three technological solutions to be compared. The established and most implemented solution, i.e. the cold air system, is taken as a *baseline* and then compared to the two alternative solutions. A comparative LCA for the competing solutions is then performed. The context parameters are fixed for the preliminary assessment with respect to a French manufacturing and assembly situation in order to identify the most relevant variables. Once this has been carried out, we build a set of scenarios to characterize the different geographical operating contexts. Finally, a new comparative LCA is performed to define the most relevant context-technology couple from these scenarios.

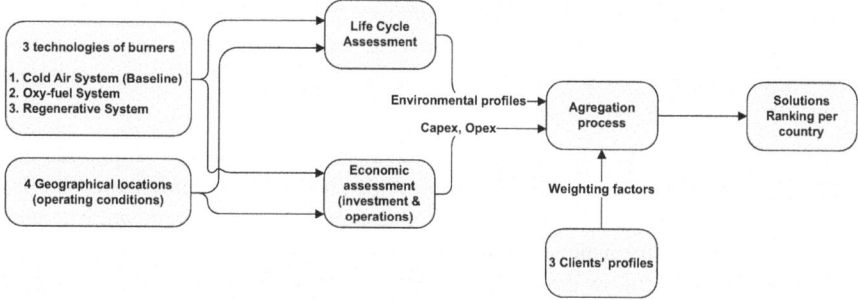

Fig. 1 Sustainable characterization and technology selection framework

Environmental issues cannot be disassociated from economic aspects. As the most environment-friendly solution may involve prohibitive costs, the decision-making process integrates the fact that the technology will be operated over a long period of time. Operational costs may also significantly contribute to the investment decision. Several works have coupled LCA with Life-Cycle-Costing (Norris 2001; Sundin et al. 2010; Widiyanto et al. 2002), which requires detailed knowledge of the system and its operation. We alternatively suggest limiting the cost integration to capital expenditure (CAPEX) and operational expenditure (OPEX). While the former focuses on investment costs, the latter includes maintenance cycles and resource consumption. CAPEX and OPEX are thus calculated and associated with each scenario.

Finally, the value system for the decision process may vary from one client to another. Thus, for the same context, the most appropriate technological solution may also vary. To overcome these limitations and shed light on the decision-making process, technological solutions are evaluated through integrating customer preferences (Masui et al. 2002). To do so, we construct realistic client profiles considering environmental and economic aspects. The three dimensions, i.e. environmental performance, economic performance and voice of customer are aggregated using several matrices that measure weighted indicators based on economic, environmental or operation parameters.

This approach is applied on three technological burner solutions for forge furnaces, considering four operating scenarios (US, France, Germany, Japan) and three customer profiles. The case study was provided by an industrial partner, a French industrial engineering group that designs and supplies process equipment, production lines and turnkey plants for the world's largest industrial groups for the aluminum, steel, glass, automotive, logistics, cement and energy sectors. The case study deals with the forging process, which involves applying compressive forces to deform steel parts. Steel parts need to be heated before being hammered, pressed or rolled. To do so, one or several burners provide uniform heat to the furnace. At present, three main alternative solutions for combustion systems are available:

- a cold air system, which consumes ambient air and fuel in the combustion process and employs proportionally greater rates of excess air as the operating rate decreases,

- an oxy-fuel system, which consumes pure oxygen, and
- a regenerative system, which recovers waste heat from furnace exhaust gases and preheats combustion air. Efficiency is significantly higher than conventional burners or burners with recovering systems.

The burner technologies were chosen as a good case study to highlight the tradeoff decision-makers have to decide on in dealing with customer expectations, geographical context (gas and electricity consumption, prices, regulations), and related industrial performances (environmental and economic).

The next sections follow the framework as depicted in Fig. 1. Methods and results from environmental assessment, economic evaluation, and the generation of the different clients are presented in the following sections. Aggregated results are then discussed before drawing conclusions and perspectives in the final section.

2 Life Cycle Assessment of a Forge Furnace

An LCA is implemented to perform environmental assessment. At present, LCA is a recognized approach to evaluate the environmental performance of product services and systems over their entire life cycle. The approach integrates life cycle stages-from the extraction of raw materials required to produce the system, to its dismantling and final waste recovery (ILCD 2010; Leroy 2009).

In this study, we compare three competing burner solutions for the forge furnace, as shown in Fig. 2, with respect to the requirements of ISO 1404X standards (International Standards Organization 2006a, 2006b).

Fig. 2 A forge furnace developed by our industrial partner

2.1 Goal and Scope of the LCA

The objective of the study is to identify the most appropriate technology to implement in a forge furnace with respect to its geographical context. The main technological choice focuses on the burner; which however cannot be disassociated from the rest of the system, i.e. the forge furnace. In this context, the system boundaries assume a cradle-to-grave perspective. This includes the raw material extraction and preparation, the manufacturing phase, the transportation phase, the operational phase (integrating the maintenance cycles) and the end-of-life treatments for both the burners and the forge furnace. The lifespan of the forge furnace is assumed to be 25 years whatever the selected technology. Several parts of the furnace, such as the refractory fibers and furnace hearth, along with the regenerative burner heat transfer media, are assumed to be replaced each year as a part of scheduled maintenance. The functional unit (FU) of the system is defined as follows: "to provide and maintain the heat (temperature of 2300°F) uniformly distributed in the furnace". The system boundaries are represented in Fig. 3.

2.2 Gathering the Data: Life Cycle Inventory (LCI)

Primary data used to populate the model were gathered from our industrial partner. Secondary data, related to competing systems and sub-systems, were extracted from the Ecoinvent V2.0 database (Frischknecht et al. 2005). The LCI is provided in Table 1. The LCI is broken down according to the different life cycle phases. Consumption (energy and raw materials) and emissions are reported for the different components, the manufacturing phase, the distribution phase, the operational phase and end-of-life management (EOL).

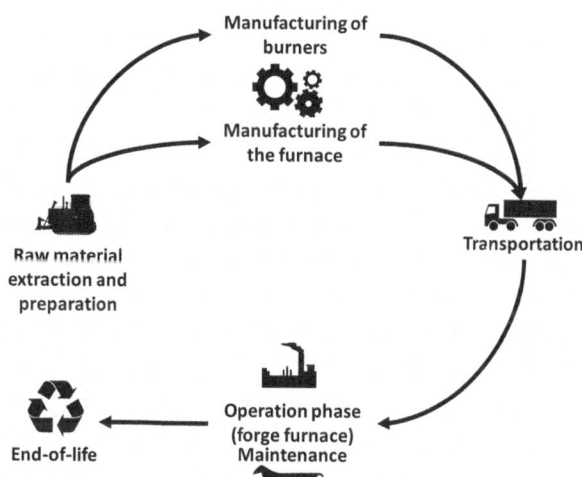

Fig. 3 Life-cycle system boundaries, a cradle-to-grave perspective

Manufacturing of burners

Manufacturing of the furnace

Transportation

Raw material extraction and preparation

Operation phase (forge furnace)
Maintenance

End-of-life

For comparison purposes, the three burner technologies are essentially designed in Europe and can be installed worldwide according to the forge furnace location. This assumption is the baseline for our case study. Thus, raw material distribution and energy consumption for the manufacturing phase are extracted from significant generic data for a European context. As the study focuses on burner technologies, the forge furnace design parameters, such as dimensions and yearly production (5500 ton/year) for example, are fixed whatever the considered burner alternatives.

2.3   Environmental Assessment, Life Cycle Impact Assessment (LCIA)

The LCIA was performed using Simapro 7.2 software. As previously mentioned, most secondary data were extracted from the Ecoinvent V2.0 database. Environmental impacts are assessed using the CML 2000 V2 characterization method (Guinée et al. 2002). The ten environmental impact categories are: monitored abiotic depletion, acidification, eutrophication, global warming, ozone layer depletion, human toxicity, fresh water aquatic ecotoxicity, marine aquatic ecotoxicity, terrestrial ecotoxicity, and photochemical oxidation.

Most of the data related to raw material consumption and emissions are reported in Table 1. Note several data are missing and marked as '?'. This case essentially occurs when dealing with EOL strategies. For each of geographical scenario, the '?' were replaced with the current rates applied per country.

3   Results of the Environmental Assessment

From an environmental viewpoint, the best solution seems to be the regenerative system, which ranks first for most impact categories (except for fresh water aquatic ecotox). The following position is shared between the two other systems, i.e. the cold air system and the oxy-combustion. Results are presented in Fig. 4.

Once the global environmental impact is broken down according to life cycle phases, the three eco-profiles appear to be quite similar. Indeed, the operational phase is responsible for the main share of environmental impact (see Table 2) as reported in previous LCA studies on energy-consuming industrial systems (Cluzel et al. 2014). The other life cycle phases often make a lower contribution. However, several impact categories are more sensitive to these phases, especially in the case of manufacturing and maintenance. Apart from ozone layer depletion, the categories of interest deal with toxicity (human toxicity, eco-toxicity, fresh water eco-toxicity). All of these significant impacts mainly occur because of raw material extraction and preparation.

Table 1 Life cycle inventory

		Material	Unit	Cold air system	Oxy-combustion system	Regenerative system
Components	Furnace	Steel	kg	18,665.33	18,665.33	18,665.33
		Refractory (roof and walls)	kg	3175.15	3175.15	3175.15
		Refractory (bottom part of walls)	kg	19,368.39	19,368.39	19,368.39
		Refractory (furnace hearth)	kg	204.12	204.12	204.12
	Burners	Refractory (chimney)	kg	2721.55	2721.55	2721.55
		Steel	kg	122.47	122.47	1197.48
		Controls panel	kg	9.07	9.07	9.07
		Computers	Unit	10.00	10.00	10.00
		Refractory tile	kg	40.82	40.82	–
		Refractory	kg	–	–	1678.29
		Refractory fibers	kg	–	–	18.14
		Regenerative media	kg	–	–	675.85
Manufacturing	Furnace	Welding of furnace	m	228.59	228.59	228.59
		Refractory assembly	btu	18,000,000.00	18,000,000.00	18,000,000.00
		Wastewater treatment	cu ft	35.00	35.00	35.00
	Burners	Welding of burners	m	–	–	50.00
		Refractory assembly	gal (US liq)	–	–	55.00
Transportation	Road		tkm	165,000.00	165,000.00	165,000.00
Operation (1 year)	Energy consumption (year)	Electricity for fan	kg	48,000.00	0.00	79,740.00
		Fuel	btu	18,200,000,000.00	10,920,000,000.00	10,920,000,000.00
		Control panel	kWh	28,908.00	28,908.00	28,908.00
	Consumption (year)	Oxygen	kg	–	816,466.27	–

(continued)

Table 1 (continued)

	Material	Unit	Cold air system	Oxy-combustion system	Regenerative system
Emissions into the air (year)	Nox	kg	1238.31	45.36	306.17
	CO	kg	82.55	82.55	90.72
	CO2	t	1066.00	619.00	555.00
Maintenance (year)	Refractory fibers	kg	1058.38	1058.38	1058.38
	Furnace hearth	kg	19,050.88	19,050.88	19,050.88
	Regenerative media	kg	–	–	181.44
End-of-life					
Furnace	Steel	kg	?	?	?
	Refractory fibers	kg	?	?	?
	Refractory furnace hearth	kg	?	?	?
Burner	Steel	kg	?	?	?
	Refractory	kg	?	?	?
	Regenerative media	kg	–	–	?

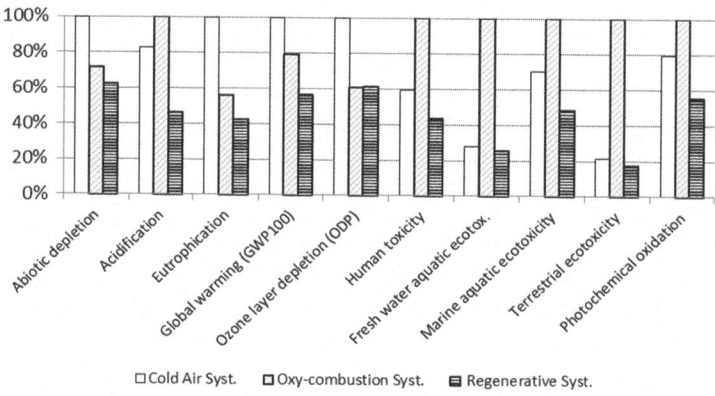

Fig. 4 Comparative LCA of three burner technologies (US Location)

Table 2 Environmental contribution of forge furnace per life cycle phases (United States location)

Life cycle phases	Cold air syst.	Oxy-combustion syst.	Regenerative syst.
Manufacturing (%)	0.1–15.5	0.2–4.6	0.2–15.2
Transportation (%)	0.1–0.8	0.1–1.0	0.1–1.3
Operation (%)	92.4–99.9	97.7–99.8	85.4–99.7
EOL (%)	−8.5−−0.1	−2.7−−0.1	−1.2−−0.1

The same situation is observed with the maintenance phase, which substitutes specific components that require consuming additional resources. In addition, the impacts of the operational phase are essentially due to electricity consumption (fan and oxygen production) and natural gas consumption (to feed the burners). Impacts related to the EOL are quite low despite the annual maintenance cycle for furnace components. This can be explained by the simplified EOL scenario (steel collection and recycling) where most of materials collected are assumed to be sent to landfill.

4 Scenario Modeling

According to the LCA results, energy and the resource consumption observed in the operational phase are the one that impact the elements of the life cycle the most. Operational costs associated with this resource consumption (electricity in particular) may also vary from one country to another. By way of consequence and in order to capture the spatial variability, we illustrate this study by choosing four countries with different locations (United States, France, Germany and Japan), energy mixes, natural gas supplies and EOL performances. Considering these four countries and the three alternative burner technologies described above, twelve scenarios are thus considered in the following analyses. As mentioned in the introduction, the economic aspect is considered in these scenarios.

5 Economic Assessment

As costs are of high interest in the decision-making process, cost assessment is performed to characterize each scenario, i.e. each solution operating in a given geographical context. To do so, Capital Expenditure (CAPEX) and Operational Expenditure (OPEX) are considered. CAPEX is similar for all context, while OPEX is sensitive to the market prices of specific resources and resource availability as reported in Fig. 5. An example for the US is given in Table 3. The same process is implemented for the three other locations (France, Japan and Germany). We consider the average American energy mix in this part. The pricing is based on 2012 US prices for electricity, gas and liquid oxygen in the industrial sector (Eurostat and IEA 2013). In the simulations, energy mixes and prices are assumed to be constant over the forge's expected lifespan (25 years).

With the investment costs provided by our industrial partner, which were size dependent, we chose an average value. We did not however take into account the cost of maintenance. Operating costs exclusively integrate the cost of operations and purchase of resources (electricity, gas, oxygen).

Taking the results in Table 3, if used in the same context (without any specific technical requirements, e.g. higher temperature, regulatory limitations, or NOx limitations) the Oxy-combustion system would not be selected. Indeed, the operational costs are 10 times higher than for the other two technologies.

The choice between cold air and regenerative technologies comes down to deciding between higher investment costs or lower total life-cycle costs. The return on over-investment for regenerative over cold air (after which time the total cost of the regenerative system becomes cheaper) is almost 5.8 years (Table 3). Considering the entire life cycle, the regenerative system is 16% cheaper than the cold air system.

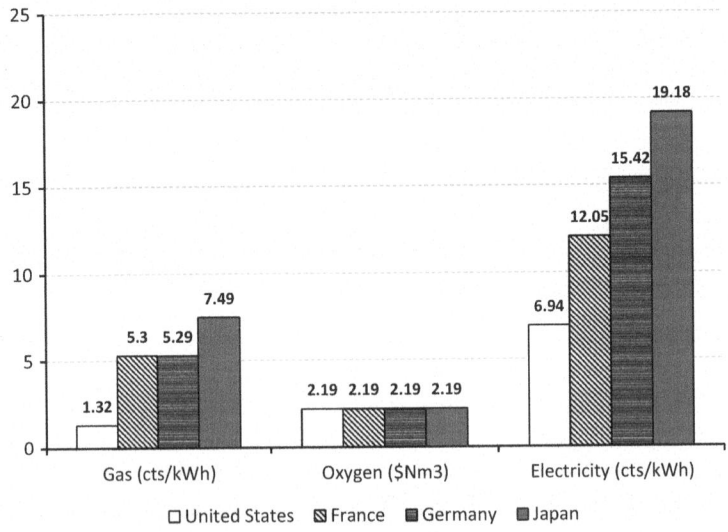

Fig. 5 Prices of energy sources according to operational location

Table 3 Cost breakdown for the three burner technologies (US location)

	Components	Unit	Cold air system	Oxy-combustion system	Regenerative system
Investment	Furnace	US$	1,000,000	1,000,000	1,000,000
	Burners	US$	150,000	100,000	300,000
	Total	US$	1,100,000	1,100,000	1,400,000
Operating costs (per year)	Gas	US$	70,407	42,244	42,244
	Oxygen	US$	–	1,251,191	–
	Electricity	US$	5337	2006	7540
	Total/year	US$	75,745	1,295,441	49,785
Total cost (25 years)		US$	3,043,620	33,486,036	2,544,615

However, over this length of time, uncertainties regarding energy and gas prices are high. At a later date, for scenario building and analysis, oxygen prices are the same for all locations due to the lack of available and relevant price data.

We conducted a similar assessment for the three other locations (France, Germany, and Japan). The results are quite similar but are not detailed in this chapter.

6 Generation of Customer Profiles

As clients may change their preferences according to their own system of values, the most relevant technological solution may be different according to an individual client's profile. In order to integrate such a dimension into the framework, we built three client profiles and their own system of values. For each of them, economic and environmental aspects are ranked and weighted according to their value perception. The following customer profiles are presented in Table 4:

– Client #1: Short-term economic vision, no environmental consideration;
– Client #2: Long-term economic vision, environmental awareness;
– Client #3: Long-term economic vision, environmental champion.

Each score is weighted on a maturity scale from 0 to 10, with 0 corresponding to 'no consideration' and 10 to 'high consideration'. With respect to the aforementioned

Table 4 Customer profiles considered

Variables	Client #1	Client #2	Client #3
Capex	10	10	10
Opex	5	10	10
Global warming potential	2	5	10
Ozone layer depletion	0	2	5
Human toxicity	0	2	5
Fossil depletion	0	2	5
Recycling/reuse	0	5	10

results, the environmental impacts are essentially due to gas consumption, electricity (fan and oxygen production) and natural gas in all cases. They mostly contribute to global warming (electricity production and CO, CO_2 combustion emissions), ozone layer depletion (electricity and gas production), human toxicity (electricity production and NOx emissions) and fossil depletion (gas production). For these reasons, the number of impact categories of interest is reduced to these last four impact categories. Raw material consumption is also of high interest when considering maintenance cycles. In order to integrate such an issue, we develop a new parameter, so-called recycling/reuse. This takes into consideration the average recycling rate for each country (Steel Recycling Institute 2013; World Steel Association n.d.). According to industrial data, only steel is recovered when refractory parts are discarded. Maintenance cycles are assessed through both fossil depletion and recycling/reuse indicators.

Once populated, customer expectations are used to weight the economic and environmental results obtained from the previous simulations. From this process, a comprehensive ranking of the three technological solutions is obtained.

7 Results

Economic and environmental simulations are applied for the United States, Japan, France and Germany. These geographical locations were chosen because of their specific energy mixes (electricity production), the specific accessibility to natural gas and their specific End-of-Life strategies. The seven variables; namely Capex, Opex, global warming potential, ozone layer depletion, human toxicity, fossil depletion and recycling/reuse, are evaluated for the three customer profiles and reported in Table 5. Results are expressed in relative values compared to the cold air system, which is

Table 5 Detailed results for the regenerative system (Client'#3)

Rate of importance	Weight	Variables	Unit	France	Japan	United States	Germany
18.2%	10	Capex	US$	0.050	0.050	0.050	0.050
18.2%	10	Opex	US$	−0.068	−0.067	−0.062	−0.067
18.2%	10	Global warming potential	kg CO_2 eq	−0.078	−0.077	−0.076	−0.077
9.1%	5	Ozone layer depletion	kg CFC-11 eq	−0.035	−0.019	−0.035	−0.035
9.1%	5	Human toxicity	kg 1,4-DB eq	−0.022	−0.025	−0.015	−0.025
9.1%	5	Abiotic depletion	kg Sb eq	−0.034	−0.034	−0.033	−0.033
18.2%	10	Recycling	kg	−0.225	−0.225	−0.225	−0.225
100.0%				−0.412	−0.397	−0.397	−0.412

taken as the baseline. Once normalized, results are then weighted with respect to client preferences. An aggregated score is then calculated as the weighted sum of these variables (expressed as single values), in order to sort and rank the three technological solutions.

Observing the results for customer profile 3, long-term economic vision and environmental champion, economic issues (CAPEX and OPEX), global warming potential and EOL (recycling rate) are of high importance in decision-making (18.2%). The three other supervised variables, namely ozone depletion, human toxicity and abiotic depletion, are of minor importance (9.1%). A slight difference is observed for the CAPEX between the forge furnace equipped with the baseline and the forge furnace equipped with the regenerative system, and the value obtained is equal to 0.05. In contrast, the OPEX values are quite different. These negative values present lower operational costs for the regenerative solution. Indeed, lower amounts of fuel and electricity are consumed during the 25-year operation phase. Results are similar regardless of the specific location. Focusing on environmental impact categories, all values are negative. This suggests that the regenerative system has higher environmental performances. This solution presents better ozone depletion value in the Japanese context. The main reason is the specific natural gas supply and the energy mix, which generate less impact than for the other locations. The recycling variable also supports the regenerative system. Although the quantity of raw materials consumed is higher for this solution (construction and maintenance), the environmental profile is balanced thanks to a higher amount of recyclable materials. Finally, an aggregated value is calculated for each country in order to overcome complex decision-making situations. In the example reported in Table 5, the regenerative solution is the preferred alternative whatever the geographical location. The aggregate indicator is used to allow the decision-maker to combine all variables and shed light on the decision.

Figure 6 provides additional results combining the specific location, the economic and environmental performances, and the client profiles. As previously mentioned, the results are expressed in relative values and are normalized compared to the cold air system (baseline). The tornado diagrams report the overall performance (aggregated value) of each technological solution per client expectation and per country. Thus, if values are positive, the cold air system is defined as the most relevant solution. On the contrary, in case of negative values, the alternative solution is preferable. In addition, a comparison between the regenerative system and the oxy-combustion system is possible. The three client profiles integrate the specific values system (Table 4) and position the technological solutions according to their expectations.

Results for all countries lead to similar conclusions. The ranking of the three technological solutions favors the implementation of the regenerative system with respect to client profiles 2 and 3, integrating both environmental considerations into the decision-making process. In this context, the Oxy-combustion System is perceived as the worst solution. On the other hand, client profile 1 points to a preference for the cold air system, although results compared to the regenerative solution are very close. Indeed, the regenerative solution presents lower gas consumption during the operational phase than the baseline does. Considering the Oxy-combustion system, the main additional impact comes from liquid oxygen consumption, which requires significant energy to be produced. Although the profiles are similar from one country to another, the slight differences observed are linked to the energy prices. Three trends can indeed be observed: the US, Western European and Japanese situations respectively. France and Germany provide an intermediate profile. The US strongly supports the regenerative system and the cold air system compared to the oxy-fuel system. Results for Japan are closer and the superiority of both alternatives compared to the oxy-fuel system is attenuated. The main explanations are global environmental impacts and energy costs (production and transportation). Japan produces the highest environmental impacts for the three technologies among the four countries. The economic issue reveals the same trends while observing the costs of gas and electricity as reported in Fig. 5. In the case of client profile 1, i.e. without any environmental consideration, the regenerative system appears to compete with the baseline. The consideration of environmental impact categories, i.e. profiles 2 and 3, decreases the performances of the two alternative technological solutions. In consequence, the more environmental issues that are integrated into the decision-making, the more attractive the regenerative system becomes.

8   Discussion

In this chapter, we investigate the performances of three technological burner solutions for a forge furnace. Environmental profiles are drawn thanks to LCA, while economic issues are dealt with through calculating the CAPEX and OPEX. The results highlight several issues when dealing with decision-making under sustainable considerations. First, comparative LCAs of technological solutions are often performed considering generic models and databases. Thus, variability caused by local situations is erased allowing for a ranking of solutions to be easily obtained. Nevertheless, local context may significantly affect environmental performance. It is quite important to consider local variables up to the point when they integrate resource availability and accessibility, local or regional energy mixes, and specific production processes. Consequently, there is a real need for detailed analysis to overcome simplifications in generic models for this type of application. Second, sustainable decisions cannot focus exclusively on the environmental dimension. Economic issues and client expectations are of great importance in the decision-making process (Norris 2001). The simple integration of both CAPEX and OPEX provides useful information to the decision-maker in a life cycle approach while long-term and short-term visions can be handled. Coupling such economic and environmental results with customer expectations facilitates technological choices. While some clients are sensitive to environmental considerations, others are not. Knowledge of customer preferences can shed light on decision-making by identifying the most relevant and suitable solution. Results from Fig. 6 support this fact. Indeed, in all countries, the most relevant solution for client 1 is the cold air system

Fig. 6 Simulation results for the three client profiles, the three alternatives systems and the four locations. The cold air system is taken as reference

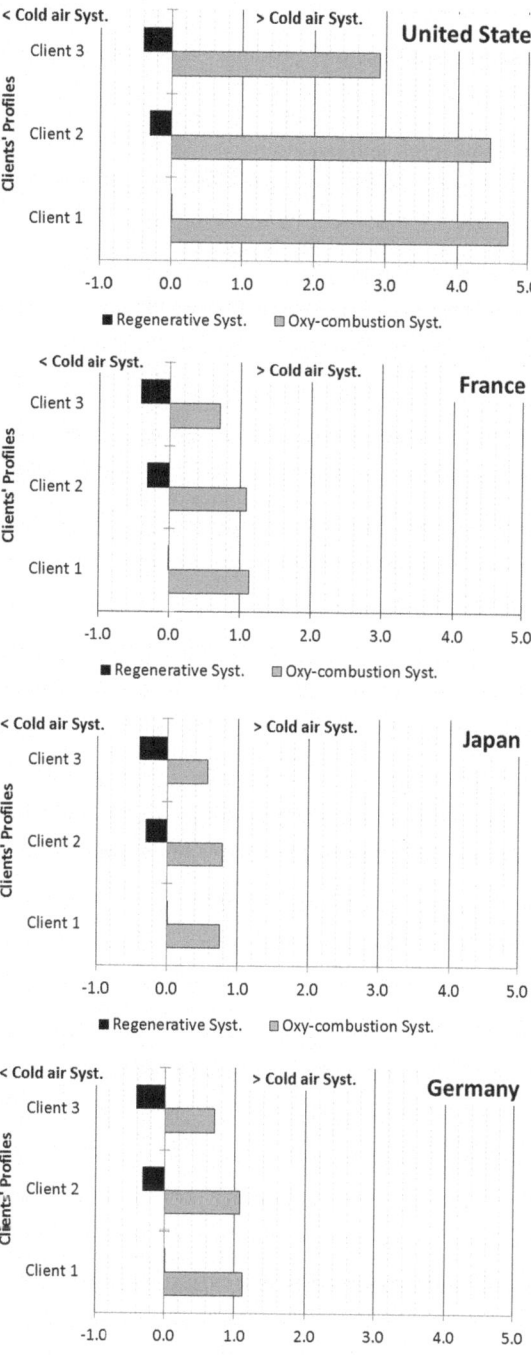

(i.e. the baseline solution). The increasing integration of environmental aspects in client profiles 2 and 3 produces a new ranking. We can thus observe a reversal between the baseline and the regenerative system.

The modeling approach also raises several issues and limitations. First, both environmental and economic modeling and simulations are static. The study excludes the market price evolution for gas and electricity, the energy mix composition, and demand for raw materials that may highly impact operational costs. In addition, it excludes the technological evolutions of local production processes, which may contribute towards reducing environmental burdens. The same assumption was also made regarding the regulatory context. These limitations pave the way for avenues of further improvement towards more robust simulations.

Limitations can be observed on data availability and data relevancy. Indeed, LCA requires a great deal of information. Most of this information is extracted from LCI databases, which provide uncertain data completeness and relevancy (Björklund 2002; Leroy 2009). In the case study, particular attention was paid to data collection and data selection. However, data on the oxy-combustion system remains lacking. One example is the price of such technology. In order to account for the lack of knowledge, the same price was assumed for liquid oxygen whatever the country considered.

Finally, the methodology is based on a multi-criteria analysis. LCA may focus on a single impact category, such as the Global Warming Potential, or more than one. The number of impact categories was reduced from ten (Fig. 4) to four (Fig. 6) in the case study. This simplification was performed in order to ease the interpretation. These categories were chosen with respect to the sensitivity analysis. Nevertheless, this limitation indicates a good way to enrich the case study in future work.

Furthermore, the ranking of the technological solution is based on weighting factors extracted from client expectations. Knowing that a multi-criteria analysis can lead to an impossible decision, an aggregated score was also calculated consisting in the sum of economic and environmental weighted values. Results could be consolidated through implementing more consistent multi-criteria approaches (Wang et al. 2009). This would increase the robustness of such a decision-making process.

Finally, another direction for improvement could be the consideration of uncertainty intervals, as close results are often not sufficient to make a reliable decision. Several quantitative or qualitative uncertainty methods, such as intervals propagation, Monte Carlo simulations, and the Bayesian Monte Carlo Approach, could be easily implemented to enrich the proposed approach (Lo et al. 2005; Lloyd and Ries 2007; Leroy and Froelich 2010).

9   Conclusions and Perspectives

We have proposed in this chapter a framework to assess the sustainable performance of industrial systems by including environmental, economic, commercial and local aspects. This approach is based on LCAs of different technological alternatives. A first LCA is performed to identify the most relevant parameters and thus to generate a set of operation scenarios associated with different geographical locations. Once the environmental profiles of these scenarios are identified, the economic costs, based on OPEX and CAPEX, are associated. The consideration of different customer profiles, corresponding to different industrial positions, then permits a relative and weighted performance of an alternative solution compared to a predefined reference, for the selected indicators (i.e. costs or environmental impact categories). The analysis of these indicators may be sufficient for decision-making; however for some configurations it proves impossible. That is why we propose an aggregate indicator that may be useful to support decision-making.

Applied to burner technologies for forge furnaces, this framework shows that local context data are essential to assess the sustainability of industrial systems and make reliable decisions. Indeed, in some cases, the use of generic data may lead to false results and wrong decisions.

Although this model could be improved, for example by refining and completing data, it clearly identifies the need for carefully defining a relevant context when assessing the sustainability of industrial systems. Future research could consider other industrial case studies, or take into account more structured uncertainty or decision-making methods and tools to propose an improved sustainability assessment framework.

References

Bertoluci, G., Leroy, Y., & Olsson, A. (2014). Exploring the environmental impacts of olive packaging solutions for the European food market. *Journal of Cleaner Production, 64.* https://doi.org/10.1016/j.jclepro.2013.09.029.

Björklund, A. E. (2002). Survey of approaches to improve reliability in lca. *International Journal of Life Cycle Assessment, 7.* https://doi.org/10.1007/BF02978849.

Cluzel, F., Yannou, B., Millet, D., & Leroy, Y. (2014). Exploitation scenarios in industrial system LCA. *International Journal of Life Cycle Assessment, 19.* https://doi.org/10.1007/s11367-013-0631-z.

Eurostat & IEA. (2013). *Industrial electricity prices in the EU and G7 countries* [WWW Document]. https://www.gov.uk/government/statistical-data-sets/international-industrial-energy-prices

Frischknecht, R., Jungbluth, N., Althaus, H., Doka, G., Dones, R., Heck, T., Hellweg, S., Hischier, R., Nemecek, T., Rebitzer, G., & Spielmann, M. (2005). Ecoinvent: Introduction the ecoinvent database: Overview and methodological framework. *International Journal of Life Cycle Assessment, 10,* 3–9.

Guinée, J., Gorrée, M., Heijungs, R., Huppes, G., Kleijn, R., van Oers, L., Wegener Sleeswijk, A., Suh, S., Udo de Haes, H., de Bruijn, H., van Duin, R., & Huijbregts, M. (2002). *Life cycle assessment. An operational guide to the ISO standards.* Dordrecht: Kluwer Academic Publishers.

ILCD. (2010). *ILCD handbook, International reference life cycle data system – Framework and requirements for life cycle impact assessment models and indicators.* European Commission. https://doi.org/10.2788/38719.

International Standards Organization. (2006a). *ISO 14040: Environmental management – Life cycle assessment – Principles and framework.* ISO 14040:2006(E).

International Standards Organization. (2006b). *ISO 14044: Environmental management – Life cycle assessment – Requirements and guidelines.* ISO 14044:2006(E).

Leroy, Y. (2009). *Development of a methodology to reliable environmental decision from life cycle assessment based on analysis and management of uncertainty on inventory data*. Ecole Nationale Supérieure des Arts et Métiers.

Leroy, Y., & Froelich, D. (2010). Qualitative and quantitative approaches dealing with uncertainty in life cycle assessment (LCA) of complex systems: Towards a selective integration of uncertainty according to LCA objectives. *International Journal of Design Engineering, 3.* https://doi.org/10.1504/IJDE.2010.034862.

Lloyd, S. M., & Ries, R. (2007). Characterizing, propagating, and analyzing uncertainty in life-cycle assessment: A survey of quantitative approaches. *Journal of Industrial Ecology, 11,* 161–179.

Lo, S., Ma, H., & Lo, S. (2005). Quantifying and reducing uncertainty in life cycle assessment using the Bayesian Monte Carlo method. *Science of the Total Environment, 340.* https://doi.org/10.1016/j.scitotenv.2004.08.020.

Masui, K., Sakao, T., Aizawa, S., & Inaba, A. (2002). Quality function deployment for environment (QFDE) to support Design for Environment (DFE), in: Volume 3: 7th Design for Manufacturing Conference. *ASME,* 415–423. https://doi.org/10.1115/DETC2002/DFM-34199.

Norris, G. A. (2001). Integrating economic analysis into LCA. *Environmental Quality Management, 10.* https://doi.org/10.1002/tqem.1006.abs.

Reap, J., Roman, F., Duncan, S., & Bras, B. (2008a). A survey of unresolved problems in life cycle assessment. Part 1: Goal and scope and inventory analysis. *International Journal of Life Cycle Assessment, 13.* https://doi.org/10.1007/s11367-008-0008-x.

Reap, J., Roman, F., Duncan, S., & Bras, B. (2008b). A survey of unresolved problems in life cycle assessment. Part 2: Impact assessment and interpretation. *International Journal of Life Cycle Assessment, 13.* https://doi.org/10.1007/s11367-008-0009-9.

Smith, A. D., & Mago, P. J. (2013). *Economic, emissions, and energy benefits from combined heat and power systems by location in the united states based on system component efficiencies.* In: Volume 6A: Energy. ASME, San Diego, p. V06AT07A034. https://doi.org/10.1115/IMECE2013-64566.

Steel Recycling Institute. (2013). *Steel recycling rates in the US* [WWW Document]. http://www.recycle-steel.org

Sundin, E., Lindahl, M., & Larsson, H. (2010). *Environmental and economic benefits of industrial product/service systems.* Proc. CIRP Industrial Production System.

Wang, J.-J., Jing, Y.-Y., Zhang, C.-F., & Zhao, J.-H. (2009). Review on multi-criteria decision analysis aid in sustainable energy decision-making. *Renewable and Sustainable Energy Reviews, 13.* https://doi.org/10.1016/j.rser.2009.06.021.

Widiyanto, A., Kato, S., & Maruyama, N. (2002). A LCA/LCC optimized selection of power plant system with additional facilities options. *Journal of Energy Resources Technology, 124,* 290. https://doi.org/10.1115/1.1507329.

World Steel Association. (n.d.) *No Titl* [WWW Document]. http://www.worldsteel.org/

Part III
The Regional Approach to Sustainable Transport as a New Paradigm

Smart Mobility Providing Smart Cities

Isabelle Nicolaï and Rémy Le Boennec

Abstract By 2050, 70% of the world's population will live in or around a city. Cities already generate 70% of energy-related greenhouse gas emissions. The future of urbanisation will be smart, in which land use is optimised and the transport system is more efficient and environmentally friendly, providing affordable mobility services to ensure well-being in the city.

In a smart city, urban and transport planning should be co-conducted harmoniously in order to create a new transit-supportive city, which is the wider context in which we position our vision of smart mobility. After this we present and analyse the links between the transport system, disruptive innovation, and the role of public policies in change management. In this chapter, we focus on the organisation of the co-conception of smart mobility, in a local territory, defining this as disruptive eco-innovation. The development and diffusion of innovations within the mobility ecosystem significantly disrupt usages and modify market boundaries. Implementation conditions to achieve a widespread adoption of smart mobility are discussed and the role and decision-making methods of territorial actors are considered.

Keywords Smart mobility · Smart city · Disruptive innovation · Eco-innovation · Territory · Public policy · Governance

I. Nicolaï (✉)
Laboratoire Genie Industriel, CentraleSupélec, Université Paris Saclay, Gif sur Yvette, France
e-mail: isabelle.nicolai@centralesupelec.fr

R. Le Boennec
Laboratoire Genie Industriel, CentraleSupélec, Université Paris-Saclay, Gif-sur-Yvette, France

VEDECOM Institute, Versailles, France
e-mail: remy.leboennec@vedecom.fr

© Springer International Publishing AG, part of Springer Nature 2018
P. da Costa, D. Attias (eds.), *Towards a Sustainable Economy*,
Sustainability and Innovation, https://doi.org/10.1007/978-3-319-79060-2_7

1   Introduction: The Lack of Transport Systems in the Least Dense Territories

In order to work or study, to consume and to build relationships, people have a fundamental need to travel. The mobility of people and goods strengthens exchanges and enables access to much larger territories (Didier and Prud'homme 2007; Le Boennec 2013). People move in space: within cities, from the suburbs to the city, to the country, between two cities, two regions or two countries. What mobility has in common in these heterogeneous spaces is that it provides people with better accessibility to jobs, goods and services. Accessibility is thus defined as the number of travel opportunities for a given distance (Pouyanne 2005) or for a given travel time (Deymier 2007). Accessibility gains occur when a new transport infrastructure enables the individual to reach the same destinations in a shorter time, or destinations further away with no increase in time, thus affording new travel opportunities.

Accessibility has extremely different characteristics depending on the territory considered, influencing the modal decisions of households. In 2013 in Paris, France, 100% of inhabitants were served by structuring transport accessible at less than 1 km from their home (Marks 2016). This privileged situation is correlated with a high population density coupled with a public transport network that is historically well connected (particularly the metropolitan network). If one considers the inner circle of Paris (the three departments around the city centre), then the outer Parisian circle (composed of the remaining territories of the Ile-de-France Region), this level of cover of inhabitants by structuring transport falls dramatically to 66%, then only 13%, respectively.

This heterogeneous accessibility to a public transport network within the same urban area[1] constitutes a direct manifestation of the rapidly decreasing density gradients from the city centre. It is much easier to connect a territory effectively where the housing density, and thus the population, is high (generally, the central municipality of an urban unit). Conversely, in less dense zones, a transport network, especially structuring (underground, tram, train or bus rapid transit), necessarily favours certain zones, with some irreversibility due to the size of the infrastructure deployed. This less favourable situation usually corresponds to municipalities that are peripheral to the urban unit, or to municipalities of the urban area around the urban unit (commonly called the peri-urban space). Regarding rural municipalities, they are rarely served other than by a bus network of varying size and frequency.

Such heterogeneity in public transport provision can be seen in most European conurbations, where collective housing is largely concentrated in the urban unit; individual housing is more often located in the peripheral municipalities, on building

---

[1]An urban area in France is a group of contiguous municipalities, with no pockets of clear land, constituted by an urban centre (or urban unit) providing more than 10,000 jobs, and by rural municipalities or urban units (urban periphery) in which at least 40% of the employed resident population works in the centre or in the municipalities attracted by it (French national statistics institute *INSEE* definition).

plots whose size is positively correlated with their distance from the centre, consistent with the urban economics theory (Alonso 1964; Fujita 1989; Le Boennec 2014). In emerging countries, very rapid urbanisation, often poorly controlled, does not always allow for sufficiently thought out coordination between urban planning and transport systems (Marks 2016).

It is thus the urban sprawl that is at the origin of the contrasting accessibility to the public transport network. This sprawl has only one consequence: in the less dense housing areas, peripheral municipalities of urban areas and rural municipalities, the private car is often the only solution for daily journeys (Brownstone and Golob 2009).[2] Thus, the car is used daily by 10% of Paris inhabitants, 26% of those in the inner circle, but 56% of the outer circle inhabitants (source: Chronos/L'ObSoCo study, "L'Observatoire des mobilités émergentes", The Observatory of Emergent Mobilities, September 2016). Similar travel patterns are found in the rest of France; while 54% of city centre inhabitants think that they can do without their (privately-owned) car completely, this can be envisaged by only 27% of people in conurbations of 2–20,000 inhabitants. As a result, in all territories taken together, 85% of journeys in 2016 were still undertaken in private cars.

This contrasting accessibility, and the heterogeneity of opportunities that it offers, is often, in an urban environment, the legacy of functionalist urban planning between the 1950s and the 1970s, at least in France. Ways of living, working and consuming often concern separate spaces, which may be far from each other even within the same urban unit. Moreover, the aspiration to own an individual house for a large number of households, coupled with easier home ownership, has increased land use and the urban sprawl beyond the urban unit. In these peri-urban territories, accessibility is limited, including by private car in rural municipalities, with access times to major services (leisure, studies, health) systematically longer than the average (source: Insee Première n° 1579, January 2016).

In these patterns of development, in which the private car dominates, the negative externalities associated with an exclusive use of the automobile are reinforced. These external effects are known: traffic noise, greenhouse gas (GHG) emissions, and air pollution (Verhoef 1996). Although most are suffered relatively little by the inhabitants of less dense zones, GHG emissions at least, due to their worldwide dimension, constitute a universal problem that public policies must take into account in their mobility and urban development plans.

We will begin by considering the difficulty of changing local planning policies for a better coordination with local mobility policies (Sect. 2). We will then see how, in a *smart city*, innovation enables such limitations to be, at least partly, overcome (Sect. 3). Nevertheless, deploying disruptive innovation in a territory demands the concordance of a number of conditions, the nature of which we shall consider in Sect. 4. In the context of emerging offers and more difficult decisions for public actors, we will discuss in Sect. 5 the different techniques for assessing mobility policies before concluding. Section 6 concludes the sixth chapter of this book.

---

[2]The *INSEE* typology uses the following travel motives: home to work or studies, purchasing, personal business, accompaniment, leisure or visits.

2 Reducing Inappropriate Urban Forms: Expensive and Long-Term Public Policies

There are many territorial scales required to correct the negative externalities of road transport; in fact, they are complementary. In its White Paper on Transport (2011), the European Commission drew up the "roadmap for a single European transport area—Towards a competitive and resource-efficient transport system". The states developed the European objectives into national policies (bonus-penalty systems for the purchase of vehicles, regulatory framework for trying out restricted traffic zones). On a local scale, the municipality authorities often seem relevant for the development, implementation and evaluation of local policies, whose objectives are often quite similar from one document to another,[3] but whose range of operational measures is supposed to take into account the special features of territories in connection with other planning documents concerning transportation and land-use.

In France, the constitution of metropoles on 1 January 2016 aimed to resolve certain governance problems that had been observed. The non-simultaneous implementation of multiple planning documents may have resulted in an insufficiently consistent vision of the development of a territory to the relevant boundaries (the city centre and its suburbs constituting the urban unit). The legislators have sought to change these historic practices by recently encouraging the adoption of urban plans at this scale. In 2016, the majority of the ex-urban communities that had become metropoles intended to adopt such a document. Urban planning at this scale should ensure better visibility in favour of a concerted development of housing, employment and service areas. In parallel, the mobility policies were themselves strengthened at the municipality level by the NOTRe law of August 2015. For example, since 1 January 2016, the new conurbations reaching the threshold of 50,000 inhabitants have jurisdiction over transport in their territory by becoming Autorités Organisatrices de Mobilité (AOM) (Mobility Organising Authorities). This scheme is further facilitated by the legislators who encourage the existing municipalities to come together in order to reach higher population thresholds. Thus, with some exceptions, on 1 January 2017 those new (inter-)municipalities should cover at least 15,000 inhabitants.

This rationalisation, of an unprecedented size and speed in France,[4] originating from greatly constrained budgets since 2009, has led to a reinforced integration of the governance of development and mobility policies at the local level. Moreover, the territorial projects, often pre-existing and contractualised with the higher authorities, have increased the financial means as well as the expertise available to the conurbations on the question of mobility. Thus, Transit Oriented Development contracts (TOD) have been decided jointly to serve territories of average density. The fundamental principle of such contractualisation envisages specific subsidies

---

[3]In France, Agendas 21.

[4]The number of inter-municipalities on 1 January 2017 was thus 1263 against 2062 previously, i.e. a fall of 39%.

from the higher authorities[5] for the deployment of a public transport network with a high level of service, through a commitment from the local authorities to increase the housing density in the town centres served. In addition, multimodal exchanges, park-and-ride car parks, bike lanes and footpaths conceived jointly can also be supported. In addition to financial aspects, the role of the higher authorities may include sharing and disseminating experiences of the same type, which have already been implemented in its territory or elsewhere.

The traditional cooperative modes mostly concern public transport and intermodality. However, while the developments for active modes are fairly expensive for local authorities, which may contractually agree to sharing the financing, the deployment of a public transport network with a high level of service in territories less dense than a metropolis raises the question of available funding sources, especially as the share likely to be covered by the user is substantially less than it could be in a densely urbanised environment.[6] Moreover, the timescales needed for such projects are long: while a TOD project might be on a 5-year scale, the deployment of an urban planning zone on the future site often takes from 15 to 20 years. As a result, the very gradual arrival of housing, employment and services in these zones compromises even more the profitability of the transport provided during its first years of operation.

For both these reasons (financing and long timescales), many local authorities now prefer to consider the alternative, or complementary, opportunity of introducing light mobility offers. Moreover, the idea of a Mobility Organising Authority, which has replaced that of the Transport Organising Authority in France, aims to take into account these sometimes rapid changes. The advent of Mobility Organising Authorities thus envisages "extended authority in the fields of the shared use of the automobile (car-sharing, car-pooling), active modes and urban logistics".

This promotion of the shared use of the car and multimodality offers a renewed regulatory framework for multimodal information platforms, which are part of the smart city, one of whose goals is to encourage these new mobility practices.

3 Innovation in Support of the Ecomobility Market Within the Smart City

From an innovation perspective, the transport sector, based on a combination of new energy systems and the spread of internet (typical of Rifkin's third industrial revolution (2011)) turns today toward clean vehicles and connected mobility solutions. We thus refer to sustainable mobility for all the available offers—individual or

---

[5]"Départements" and "Régions" in France.

[6]In 2010, the contribution of users to financing the public transport network in the Paris region was only 29.7%.

collective, public or private—that contribute to responding positively to economic and ecological issues in terms of reducing GHG emissions (Pillot 2011).

This trend towards a new technological paradigm applied to local territorial development is expressed in the idea of 'smart mobility', as a component of a 'smart city'.

Since the 1990s, the smart city concept has become increasingly popular in international scientific studies and national public policies have adopted this type of territorial development. There are many definitions of the smart city and we propose the following typology of its analytical models (for more details about the many evolutions of the concept, see the works of Caragliu et al. (2015)).

The first approach defines the smart city as a model based on the data needed to manage and plan the city:

> Smart Cities initiatives try to improve urban performance by using data, information and information technologies (IT) to provide more efficient services to citizens, to monitor and optimise existing infrastructure, to increase collaboration among different economic actors, and to encourage innovative business models in both the private and public sectors. (Marsal-Llacuna et al. 2014).

Consequently, the objective of a smart city is to organise its activities in order to implement and interconnect technologies, devices and services as efficiently as possible using information technologies (Hancke et al. 2013).

In this model, the smart city is conceived as an information system, which instrumentalises and interconnects its assets such as buildings, the energy or water network, and transport. According to Harrison et al. (2010), "instrumented" refers to the capability of capturing and integrating live real-world data through the use of sensors, meters, appliances, personal devices, and other similar sensors. 'Interconnected' refers to the integration of these data into a computing platform, which enables the communication of such information between the various city services.

Gradually, this very technological approach has been replaced by more open definitions of the smart city that take into account the social capital dimension and its relationships with urban development. Thus, in the second model, we have a definition of the smart city through its governance and ability to be resilient. The idea of the model is to move from a 'connected' city to a 'smart' city. 'Smart' refers to the inclusion of complex analytical, modelling, optimisation, and visualisation services in order to make better operational decisions. The model emphasises the role of human capital/education, social and relational capital, and environmental interests as important drivers of urban growth (Leydesdorff and Deakin 2011, Komninos et al. 2002).

A smart city in this perspective is one that organises the conditions for the commitment of all the stakeholders of its ecosystem in the decision process. Resilience is thus reflected by the quality of people and communities, to be connected, to manage, and to be informed (Albino et al. 2015, Repko 2012). Resilience is a factor of governance, risk assessment, knowledge and education, risk management, vulnerability reduction, and disaster preparedness and response (Baron 2012; Twigg 2009).

However, in a more operational logic with a territorial application of the smart city, it seemed necessary to propose a third model, which defines how to produce smart city strategies (Ben Leitafa 2015). To meet these objectives, we propose, in line with the studies of R. Giffinger, considering the following levers to make cities smart:

- develop new efficient services in the transport-mobility sectors; responsible housing and urbanisation; smart materials and energy networks,
- manage information systems in real and multiple time to help in decision-making (citizens, administrations, organisations),
- promote renewed governance and the financing of new services.

Based on this operational logic, we can propose a third so-called architectural approach, which breaks down the systems and dimensions of the smart city in an organic way. The literature, particularly the studies of Giffinger et al. (2007) and Dirks et al. (2010), proposes six components of the smart city that correspond to six dimensions of urban life that must be made efficient: industry, education, democracy/governance, logistics and infrastructures, efficiency and sustainability, safety and quality. These six components can be represented in Fig. 1:

In this context, we use the definition proposed by Caragliu et al. (2011): "We believe a city to be smart when investments in human and social capital and traditional (transport) and modern (ICT) communication infrastructure fuel sustainable economic growth and a high quality of life, with a wise management of natural resources, through a participated governance".

In each sector, there is a supply of and a demand for connected and smart systems with specific characteristics to be analysed in various approaches related to eco-innovation[7] (Rennings 2000; Cecere et al. 2014).

In the eco-innovative mobility system, there are many evolutions that impact the most typical drivers of eco-innovation classified as "market pull", "technology push" and "institutional factors and policy measures" (Horbach et al. 2012; Nemet 2009):

- Changing travel habits and the demand for services to increase convenience, multimodality and predictability will require mobility solutions as well as business model transformation.
- Concerning the supply, one observes, in an expanded automobile ecosystem, the integration of companies specialising in other sectors such as telecoms (Faucheux and Nicolaï 2015). Similarly, the vehicle has an evolving role beyond that of a means of travel toward a means of energy storage and production. Thus, a transport fleet of rechargeable electric vehicles enables electricity to be bought or sold on a smart grid.

---

[7]We adopt the consensual definition of eco-innovation proposed initially by Kemp and Pearson (2007): "Eco-innovation is the production, application or exploitation of a good, service, production process, organizational structure, or management or business method that is novel to the firm or user and which results, throughout its life cycle, in a reduction of environmental risk, pollution and the negative impacts of resource use (including energy use) compared to relevant alternatives". See Horbach (2016) for an overview of the eco-innovation literature.

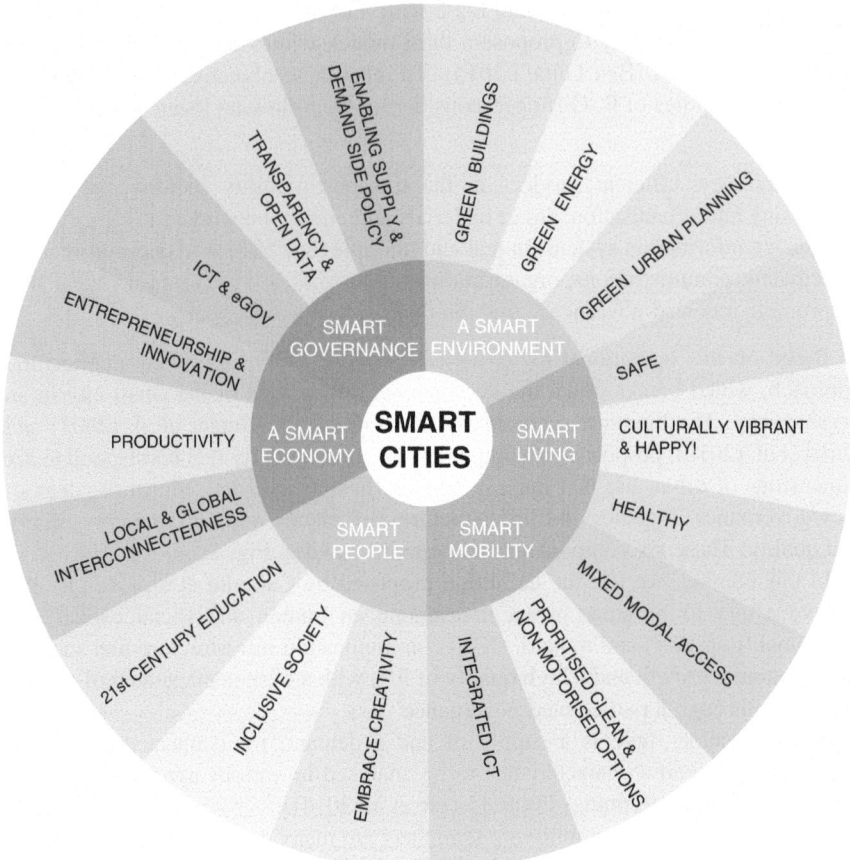

Fig. 1 The smart cities Mandala (EU & Giffinger et al.)

- A system-level approach is critical: sustainable improvements in a city's mobility performance require simultaneous improvements in urban sustainable systems, smart infrastructures, new urban services and associated business models.

In the management of a "smart mobility" system, there are additional constraints to consider. When analysing the demand for "smart mobility", we identify the following specific features of the services proposed:

1. user-focused service with an analysis of user preferences depending on the specific temporal and geographic contexts
2. a service available at every point in the territory and from all the facilities
3. an integrated service with data coming from and supplying different applications according to the principle of collaborative participation of the actors.

From the point of view of the supply of "smart mobility", the actors, such as local authorities and companies that have to be taken into account in the eco-innovation eco-system (Ghisetti et al. 2015), must offer services:

1. inter-connected within a platform that is open to heterogeneous facilities, ensuring connectivity at every point in the territory
2. with great efficiency: the organisation of such a network must be as cost-efficient as possible
3. with satisfactory energy efficiency to valorise sustainable applications
4. reliable in terms of a guaranteed connectivity in all situations, even exceptional ones.

In order to examine how the literature can transcribe the special features of 'smart mobility', we will deal with the quality of its eco-innovation to present how the new mobility solutions can be considered disruptive innovations (as defined by Christensen 1997).

4 Systemic Disruptive Eco-Innovation

4.1 A Disruptive Innovation

Ecomobility or 'smart mobility' as eco-innovation consists of both responding to the needs of consumers and encouraging their support (demand-pull innovation), while questioning which business ecosystem is most likely to provide effective solutions to these needs (technology-push innovation) in an evolving institutional context.

In this perspective, the characterisation of smart mobility according to different types of innovation is important because it will determine the conditions of its implementation.

In an ecomobility system, new needs appear and user behaviour changes in a complex co-evolution with regulatory frameworks (Dantan et al. 2017). In order to characterise the dynamics of ecomobility innovations, we will use the typology of eco-innovation proposed by Faucheux and Nicolaï (2011) in association with the different trends in the innovation economy (Utterback and Abernathy 1975, and Rennings 2000, Hellström 2007, Carrillo-Hermosilla et al. 2010).

The analysis of industrial regimes leads us to consider first the dynamics of innovation through technological trajectories as defined by Dosi (1982). In this perspective, according to Utterback (1994) innovation is identified as 'dominant design'. In the context of smart mobility, this means, for the automotive industry, imagining technologies that meet the identified needs, such as those of ownership and use of vehicles with increasing constraints or opportunities; reduction in environmental impacts, innovative economic models with, for example, the electric vehicle, taking into account recent legislation, and acceptance of the economy of functionality, which provides a mobility service instead of the acquisition of a vehicle.

We observe that over time these innovations come from a creative process that occurs in a market and production context (technology push) as well as from an adaptive response to a new demand expressed by the market (demand pull).

However, these changes go beyond the issue of a technological trajectory to promote new paradigms of smart mobility (as defined by Dosi (1982)). Thus, radical innovations (as defined by Dewar and Dutton (1986)) support technological disruptions and changes in user behaviour. This destabilisation, which was dealt with in terms of internal disruptive techniques for the overall product structure by Henderson and Clark (1990, with modular/architectural innovations), challenges the current skills of companies. The production of autonomous vehicles, as well as the development and sale of new mobility services (NMS) requires different skills for the automotive industry. However, if, like Markides (2006), we focus on the 'technological' innovation, which aims to attack new segments in an existing market[8] (a different business model in a current market), we observe that a "strategic" innovation must also be envisaged in smart mobility, which involves a different value distribution in the value chain of the automotive industry. For example, car-pooling proposes a collective use of an individual mode of transport while, conversely, car-sharing describes the individual use of a collective mode of transport. These technological innovations are applied to different uses of a traditional transport mode and transform the economic model. How can companies now promote a systemic vision of technological change, characterised by a complex structure of interactions between the socio-economic environment (markets, governments), the skills of the firm linked to technological changes, and the modifications of uses? How can they promote these systemic eco-innovations (as defined by Chesbrough and Teece (2002)) which require new internal skills and solicit new markets with few customers, sometimes in competition with their own existing market?

Smart mobility as an eco-innovation that simultaneously destabilises the supply system and the demand system (needs of users) is considered a disruptive innovation according to the definition of Mackay and Metcalfe (2002): "A disruptive innovation represents new technologies with characteristics that are initially unfamiliar to producers and consumers, which may also require a major evolution of institutions and infrastructures, and which have the potential to disrupt market structures and lead to changes in behaviours".

In smart mobility as well, we observe the following evolutions:

1. The emergence of a new technology such as electromobility (or previous automobile technologies and ICT assembled differently), which underperforms compared to the 'dominant design' (individual combustion vehicle) and the expectations of consumers. However, this new technology follows a faster technological evolution than the established one. The new mobility services (NMS) are also disruptive compared to the current use of individual and owned vehicles.

2. The technology of NMS is provided by actors at the competitive fringe, or even outside the market.[9] For example, Bolloré, a competitor outside the automobile

---

[8]The company redefines its technologies, goods and services and does not find a new market: for example Amazon, easyJet, etc.

[9]Christensen (1997) speaks of the dilemma of the innovator for whom the disruptive innovations are rarely introduced by the dominant companies in the market.

sector, has been one of the pioneers in car-sharing practices. The success of this innovation will move the strategic value of the old paradigm toward the new, which will significantly change the industrial structure, with the constitution of a new ecosystem.

3. The new technology requires new skills and a learning process by the providers to gain them. It also introduces new uses and/or considerably modifies the initial use of the technology. New practices in the economy of services related to the acquisition or maintenance of the car and the spread of connected objects in the vehicle are transformations in the uses, and thus in the supply, of products.

4. The innovations linked to ecomobility are addressed towards new users. For example, the autonomous vehicle will target the visionary 'early adopters' who are sensitive to the new technology required for its own sake. The diffusion of this innovation will follow the classic sinusoidal curve of innovation, which associates each phase of technological development with a particular category of consumers.

5. Consumers are categorised according to the time they choose to acquire the innovation in question: 'early adopter' and 'early majority'. The success or failure of the 'ecomobility' innovation will be judged when the early majority (who prefigure the transition from a niche to a mass market) adopt ecomobility with a greater importance accorded to the traditional economic determinants (price, reliability, flexibility of choice, interoperability). A new business will thus arise if the management of the transition between consumer categories is assured.

Whether the innovations in ecomobility are 'demand-pull' or 'technology-push', they can dramatically disrupt usages and change market boundaries. They thus involve new reflexions about the business model[10] that companies in the automotive industry need to develop and the evolution of the structure of the automobile ecosystem. In this sense, Christensen (1997) speaks about disruptive innovation.

Many conditions need to be fulfilled to implement these disruptive eco-innovations:

- have a design-user approach, i.e. identify and deal with what could be the uses of tomorrow in order to design technologies and products according to those needs (what are the different autonomous car models proposed by Tesla on one hand, and by existing manufacturers on the other hand?).
- agree to costly investments to explain to future users what innovation is and the value it will bring them. Given that most people are rather resistant to change and prefer to keep their habits, convincing them to adopt a new product or service that requires them to change their practices will not be a simple and rapid process. It is up to the innovative companies to identify both the facilitating and the obstructing factors to ensure a wider and faster adoption of the innovation.

---

[10]In the meaning of Christensen et al. (2002) for whom innovation corresponds to "the creation of totally new markets and economic models" (p. 22).

- invest heavily in R&D (facilities, expertise, patents) in order to stay ahead in terms of skills in the new market identified. The company must also equip itself with the new skills indispensable for developing this innovation.
- becoming the leader in a new market also involves rapid growth, i.e. continuing to invest heavily once the innovation is launched.
- lastly, the sales cycles of a disruptive innovation can be very long, certainly longer than classic innovative products. This is an additional risk factor.

The disruptive nature of smart mobility imposes a methodology of putting the user at the heart of the strategic reflexion and choosing the business model most likely to provide effective solutions to their needs in a perspective of co-conception and co-innovation. The implementation conditions are thus crucial in both the diffusion of these innovations and the success of the new business model.

4.2 Implementation Conditions

Different variables will impact whether or not the disruptive eco-innovation is successfully adopted. The first concerns the choice of developing new skills internally or not. Chesbrough and Teece (2002) envisage two organisational models to implement the innovation process: a decentralised (or virtual) approach for "autonomous"[11] innovations and an internal development of skills for "systemic" innovations. This "organisational design" is one of the accompanying variables of disruptive eco-innovations. In an institutionalist approach, it consists of identifying the most suitable institutional determinants and socio-economic organisation modes for decision-making: how do collaborative practices, known as "open innovation", promote disruptive innovation? What are the main obstacles to overcome for the public and private actors?

The second variable affects the institutional level. How can we ensure the conditions for disruptive innovations to be accepted? For example, should we preserve the acquired positions of the organisation of mobility with new actors such as *Uber*? How can we choose between the protection of private life and big data? In fact, the major pitfall for a new mobility solution is achieving the required critical mass of users; many of these solutions emerge and then just as frequently decline only 1 or 2 years after their launch. For example, this is the case of the solution *Djump* launched in 2013 but abandoned in 2015, which came close to a dynamic car-pooling. This critical mass, i.e. the transition from the adoption of the solution by the early adopters toward the early majority, is often easier to reach in dense housing areas, but this is not usually enough.

The third and last variable for a successful adoption of the offer concerns the support of public policies, with a focus on financing and territorial questions. What

---

[11]The authors speak of autonomous innovations when they can be carried out independently of each other.

are the favourable and unfavourable factors for the birth and development of the disruptive innovation? Is the financing from private and public funds sufficient and well adapted to the specific needs of the disruptive innovation, in the territories concerned? How does the territorial dimension affect the disruptive innovation? Are there "best practices" that can be transposed from one territory to another?

Most new mobility solutions are deployed primarily in central Paris, and sometimes in the Parisian inner circle and the city centres of large conurbations. This is demonstrated by the use of different private hire services *(Uber, le Cab,* etc.), whose attractiveness decreases rapidly further from the very centre of cities. Thus, while 42% of Parisian households used a private hire service at least once in 2016, only 11% did in the urban centres of more than 100,000 inhabitants outside the Ile-de-France, and only 4% in the rural municipalities (source: *Chronos/ L'ObSoCo* study 2016).

This extremely contrasting situation makes support for the development of the solution by the local authority indispensable. Even solutions that are relatively inexpensive to implement for the territories and whose success seems guaranteed, like bike-sharing schemes, remain dependent on characteristics linked to the population density, amongst others. The recent failures of this type of service in several provincial cities, some of which have more than 100,000 inhabitants (Perpignan, Angers, and Dijon), are due to various factors: too high an operational cost for the local authority compared to the revenue recovered, and under-utilisation of the system. In both cases, the lack of infrastructure or development dedicated to cyclists is highlighted; in other words, although there was sufficient public and private financing, it was not adapted to the territory.

Car-pooling is another example of a solution that may succeed or fail. While the long-distance car-pooling represented by *BlaBlaCar* has proved to be a success over the whole national territory (and internationally), the same group of solutions applied to shorter distances, like *WayzUp*, is difficult to deploy in less dense areas, despite this solution's position in the market since 2013. For a car-pooling solution to operate in more extensive territories, it often needs to be part of a local experiment: for example, the dynamic car-pooling of *Covivo* in Isère in 2010, or *RézoPouce* in the Montauban region (France). However, this experimental nature makes it difficult to identify the reproducibility of the solution, including in territories with similar socio-demographic characteristics. The local "best practices" cannot always be transposed.

To try out innovative offers, there is regular cooperation between start-ups and historic actors of mobility (car manufacturers and associated services, public transport operators). Nevertheless, these partnerships remain limited over time and are generally subject to cautious investment, thus restricted. For example, the *RATP* (Parisian public transport provider) and the start-up *Sharette* set up a partnership in summer 2015 during works on Line A of the Express Regional Network. Despite its success announced at the time, the partnership was not renewed during a new series of works in summer 2016, due to the bankruptcy of *Sharette* in the intervening period.

Promoting the socio-economic conditions for the birth and development of disruptive innovations remains insufficient. In the presence of negative externalities associated with transport, public regulatory actions are indispensable to restore the well-being of the population suffering from the negative effects to its original state. Individual mobility (solo driving) presents many advantages for the user: in a car, there are often more opportunities for travel. One can travel to destinations further away and generally in a shorter time than on public transport. The car provides "seamless mobility", door-to-door and without a break, especially in less dense areas. It is practical, offering the possibility of transporting heavy objects. Solo driving remains synonymous with independence and flexibility: the user depends on no one to make his/her journey. Lastly, it provides the driver with a space completely isolated from the outside world, if he/she wants it.

On the other hand, the use of the car leads to many negative externalities. It causes noise, air pollution, greenhouse gas emissions, traffic congestion, road accidents, etc. The role of public actors is to limit all these externalities and preserve an acceptable quality of life for the inhabitants. However, public mobility policies are often expensive to implement, even though they provide significant and long-term effects (Quinet 2010). Converting a city thoroughfare into an urban boulevard with more space dedicated to public transport, pedestrians and bicycles represents an ambitious development for a local authority, with a certain irreversibility of choice.

In this respect, mobility policies must be assessed as thoroughly and robustly as possible, before or in progress (Rousval and Bouyssou 2009). The implementation of a policy represents a number of potential projects or actions between which the decision-maker must choose with regard to various considerations. Will the project envisaged meet its objectives of correcting the negative externalities? Will it be too expensive? Is it assessed using the correct criteria? Is it preferable to another project? In other words, the cost of the opportunity of a project must be assessed.

5 Assessment Methods

In the last 15 years, less structuring mobility policies have increasingly entered the evaluation field; for example, a car-pooling centre facilitated by an average-sized conurbation (*Ministère de l'Ecologie, du Développement sustainable et de l'Energie*, 2015).

There are many economic methods to estimate the effects of a public policy. These aim to calculate the individual's willingness to pay, "that is, how much the individuals would be prepared to pay to benefit from an increase in the supply of a non-market commodity" (Meunier and Marsden 2009, p. 6). Two groups of methods are traditionally contrasted: the stated preference and the revealed preference methods (Mahieu et al. 2015). In these two groups, there are applications specific to transport projects, or to transport externalities for the local authority.

Stated preference methods are often applied to transport projects (Bristow et al. 2015). Among these, contingent valuation "estimates by survey techniques how much individuals would be ready to pay to enjoy the benefits of a project" (Meunier and Marsden 2009, p. 7). The assessment of transport projects by contingent valuation has been the subject of many applications: reduction of noise and air pollution in Navarre (Lera-López et al. 2012), estimation of parking charges in Greece (Anastasiadou et al. 2009), improvement in the provision of public transport in Dubai (Worku 2013), and rail services in Korea (Chang 2010).

In terms of revealed preferences, the best known method applied to transport projects is that of hedonic pricing. This approach is based mainly on the estimation of property prices, which depend partly on the proximity to transport networks and to sources of amenities and nuisances. Thus, the property market indirectly provides a monetary value of these attributes through the difference observed between the values of two identical properties, with the exception of one of the characteristics studied. This difference in value is explained by the gain or loss of well-being attributed by the buyers to these proximities (Rosen 1974; Le Boennec and Sari 2015). Besides the property's own characteristics and the socio-demographic variables of the neighbourhood, the variables of accessibility (to a public or road transport network) constitute another group of accommodation attributes considered in a local market. Many applications of the hedonic pricing method continue to characterise the sub-markets: in Paris (Bureau and Glachant 2010), Ireland (Mayor et al. 2012), and the United States (Duncan 2011; Bajari et al. 2012).

In these approaches, the calculation of willingness to pay may provide a basis to fuel decision-making methods of the cost-benefits analysis (CBA) type. The CBA enables the public decision-maker to choose a transport project in a context in which the decision criteria can be valued in monetary terms. In this way, even if recent evolutions have expanded the perspectives of the cost-benefits analysis, "the quantified economic assessment constitutes the core of the evaluation", as Quinet (2010) points out.

The CBA first clarifies the different elements to take into account in the analysis and provides a structured framework for the public debate (Meunier and Marsden 2009). However, several limitations of the method are generally highlighted. Some impacts are not taken into account in the assessment of projects, which risks biasing the selection; this situation arises because it is sometimes difficult to choose between an inaccurate costing of an impact, an essential condition for monetary evaluation, and no costing at all, in which case the impact will not be taken into account (Forum International des Transports 2011; Bueno et al. 2015).

Finally, although a cost-benefits analysis undeniably improves the transparency and reliability of the analysis by objectifying it, a better appreciation of the diverse effects of a mobility project may be sought by using alternative or complementary tools, in all those situations in which the analytical criteria may not easily be valued in monetary terms (Joubert et al. 1997; Tudela et al. 2006). Moreover, in practice, the CBA described above continues to favour large infrastructure projects (Beria et al. 2012; Hüging et al. 2014). For a mobility project without a very heavy infrastructure (such as bike-sharing or car-sharing), a multi-criteria analysis may be preferred by the decision-maker. Finally, it is generally agreed that a multi-criteria analysis "takes more impacts into account (...), responds more directly to the concerns of decision-makers and is open to different assessments of the weight given to different impacts" (Forum International des Transports 2011, p. 13). In many cases, this greater flexibility can make the multi-criteria analysis preferable to a cost-benefits analysis in the eyes of decision-makers (Bueno et al. 2015).

A multi-criteria analysis is implemented in several phases (Beria et al. 2012). The project or the actions to be assessed must first be specified. It is then indispensable to draw up a list of criteria that will guide the researcher in evaluating the predefined actions. The foreseeable impacts of actions should be envisaged; they must be assessed with regard to each criterion, according to a weighting system. Lastly, the aggregation of the assessments, criterion by criterion, should inform the decision-making.

The different stakeholders in the project or action, public or private, may help the researcher in the construction of the process (Beria et al. 2012). However, as described, the simple multi-criteria analysis does not explicitly take into account this diversity of stakeholders, who are not involved in the implementation stages presented above. Yet, there is a wide range of actors potentially concerned by the possible impacts of a mobility project (Macharis et al. 2009). They are primarily the users, but also the suppliers, as well as the financers (who may be partly the same but also include the public actors). This is why it may be relevant to integrate, explicitly and as soon as possible in the method, all the stakeholders in the decision-making process. Such a strategy makes the group of techniques revolving around multi-actor multi-criteria analyses (MAMCA) particularly timely (Macharis and Bernardini 2015; Le Boennec et al. 2017).

6 Conclusion

There are many challenges facing national and local mobility policies. From an environmental point of view, the fight against climate warming is a priority of local policies, which should not, however, reduce the ability of individuals to travel. From a social point of view, it is therefore a question of offering mobility solutions that meet the demands of actors to travel, specific demands that depend on the territorial contexts.

The deployment of an innovative "smart mobility" ecosystem is envisaged as an opportunity to promote a better coordination of all public transport and alternative mobility services to the private car, associated with a development of car-sharing according to different modalities.

The generalisation of ICT in ecomobility facilitates the complementarity, or even the competition or substitution, of new kinds of transport. Nevertheless, these new transport modes, developed by both public and private actors, must be coordinated to meet the new needs expressed by the demand.

What could be the role of public policies in this management of disruptive eco-innovations? Moreover, the ability to invest in and exploit infrastructures is important in the implementation of a local transport policy. In the context of a rapid evolution in mobility options and systems of governance, how could a local policy on these smart mobility issues be more effectively structured than it is today? The following chapter provides an illustration of these new responsibilities.

References

Albino, V., Berardi, U., & Dangelico, R. M. (2015). Smart cities: Definitions, dimensions, performances and initiatives. *Journal of Urban Technology*, https://doi.org/10.1080/10630732.2014.942092.

Alonso, W. (1964). *Location and land use. Toward a general theory of land rent*. Cambridge, MA: Harvard University Press.

Anastasiadou, M., Dimitriou, D. J., Fredianakis, A., Lagoudakis, E., Traxanatzi, G., & Tsagarakis, K. P. (2009). Determining the parking fee using the contingent valuation methodology. *Journal of Urban Planning and Development, 135*(3), 116–124.

Bajari, P., Fruehwirth, J. C., Kim, K. I., & Timmins, C. (2012). A rational expectations approach to hedonic price regressions with time-varying unobserved product attributes: The price of pollution. *The American Economic Review, 102*(5), 1898–1926.

Baron, M. (2012). Do we need smart cities for resilience? *Journal of Economics and Management, 10*, 32–46.

Ben Leitafa, S. (2015). How to strategize smart cities: Revealing the SMART model. *Journal of Business Research, 68*(7), 1414–1419.

Beria, P., Malteste, I., & Mariotti, I. (2012). Multicriteria versus cost benefit analysis: A comparative perspective in the assessment of sustainable mobility. *European Transport Research Review, 4*, 137–152.

Bristow, A. L., Wardman, M., & Chintakayala, V. P. K. (2015). International meta-analysis of stated preference studies of transportation noise nuisance. *Transportation, 42*(1), 71–100.

Brownstone, D., & Golob, T. F. (2009). The impact of residential density on vehicle usage and energy consumption. *Journal of Urban Economics, 65*(1), 91–98.

Bueno, P. C., Vassallo, J. M., & Cheung, K. (2015). Sustainability assessment of transport infrastructure projects: A review of existing tools and methods. *Transport Reviews, 35*(5), 622–649.

Bureau, B., & Glachant, M. (2010). Évaluation de l'impact des politiques. *Economie & prévision, 1*, 27–44.

Caragliu, A. Del Bo, C., Kourtit, K., & Nijkamp, P. (2015, December). *Smart cities in an open world*. Pre-print Politecnico di Milano.

Caragliu, A., Del Bo, C., & Nijkamp, P. (2011). Smart cities in Europe. *Journal of Urban Technology, 18*(2), 65–82.

Carrillo-Hermosilla, J., del Rio, P., & Könnölä, T. (2010). Diversity of eco-innovations: Reflections from selected case studies. *Journal of Cleaner Production, 18*, 1073–1083.

Cecere, G., Corrocher, N., Gossart, C., & Ozman, M. (2014). Lock-in and path dependence: An evolutionary approach to eco-innovations. *Journal of Evolutionary Economics, 24*(5), 1037–1065.

Chang, J. S. (2010). Estimation of option and non-use values for intercity passenger rail services. *Journal of Transport Geography, 18*(2), 259–265.

Chesbrough, H., & Teece, D. J. (2002, August). When is virtual virtuous? *Harvard Business Review*.

Christensen, C. M. (1997). *The innovator's dilemma: When new technologies cause great firms to fail*. Boston, MA: Harvard Business School Press.

Christensen, C. M., Johnson, M. W., & Rigby, D. K. (2002). Foundations for growth: How to identify and build disruptive new businesses. *Sloan Management Review, 43*(3), 22–32.

Dantan, S., Bulteau, J., & Nicolaï, I. (2017). Enhancing sustainable mobility through a multimodal platform: Would travelers pay for it? *International Journal of Sustainable Development, 20*(1/2), 33–55.

Dewar, R. D., & Dutton, J. E. (1986). The adoption of radical and incremental innovations: An empirical analysis. *Management Science, 32*(11), 1422–1433.

Deymier, G. (2007). Analyse spatio-temporelle de la capitalisation immobilière des gains d'accessibilité: l'exemple du périphérique Nord de Lyon. *Revue d'Économie Régionale & Urbaine, 4*, 755.

Didier, M., & Prud'homme, R. (2007). *Infrastructures de transport, mobilité et croissance*. Rapport du Conseil d'Analyse Economique. La Documentation Française.

Dirks, S., Gurdgiev, C., & Keeling, M. (2010). *Smarter cities for smarter growth: How cities can optimize their systems for the talent-based economy*. Somers, NY: IBM Global Business Services.

Dosi, G. (1982). Technological paradigms and technological trajectories. A suggested interpretation of the determinants and directions of technical change. *Research Policy, 11*(3), 147–162.

Duncan, M. (2011). The impact of transit-oriented development on housing prices in San Diego, CA. *Urban Studies, 48*(1), 101–127.

Faucheux, S., & Nicolaï, I. (2011). IT for green and green IT: A proposed typology of eco-innovation. *Ecological Economics, 70*, 2020–2027.

Faucheux, S., & Nicolaï, I. (2015). Business models and the diffusion of eco-innovations in the eco-mobility sector. *Society and Business Review, 10*(3), 203–222.

Forum International des Transports. (2011). *Améliorer la pratique de l'analyse coûts-bénéfices dans les transports*. Document de référence 2011–1.

Fujita, M. (1989). *Urban economic theory: Land use and city size*. Cambridge: Cambridge University Press.

Ghisetti, C., Marzucchi, A., & Montresor, S. (2015). The open eco-innovation mode. An empirical investigation of eleven European countries. *Research Policy, 44*, 1080–1093.

Giffinger, R., Fertner, C., Kramar, H., Kalasek, R., Pichler-Milanović, N., & Meijers, E. (2007). *Smart cities ranking of European medium-sized cities*. Vienna: Centre of Regional Science, Vienna University of Technology.

Hancke, G. P., Silva, B. C., & Hancke, G. P., Jr. (2013). The role of advanced sensing in smart cities. *Sensors, 13*(1), 393–425.

Harrison, C., Eckman, B., Hamilton, R., Hartswick, P., Kalagnanam, J., Paraszczak, J., & Williams, P. (2010). Foundations for smarter cities. *IBM Journal of Research and Development, 54*(4), 1–16.

Hellström, T. (2007). Dimensions of environmentally sustainable innovation: The structure of eco-innovation concepts. *Sustainable Development, 15*(3), 148–159.

Henderson, R. M., & Clark, K. B. (1990). Architectural innovation: The reconfiguration of existing product technologies and the failure of established firms. *Administrative Science Quarterly, 35*, 9–30.

Horbach, J. (2016). Empirical determinants of eco-innovation in European countries using the community innovation survey. *Environmental Innovation and Societal Transitions, 16*, 1–14.

Horbach, J., Rammer, C., & Rennings, K. (2012). Determinants of eco-innovations by type of environmental impact—The role of regulatory push/pull,technology push and market pull. *Ecological Economics, 78*, 112–122.

Hüging, H., Glensor, K., & Lah, O. (2014). Need for a holistic assessment of urban mobility measures – Review of existing methods and design of a simplified approach. *Transportation Research Procedia, 4*, 3–13.

Joubert, A. R., Leiman, A., de Klerk, H. M., Katua, S., & Aggenbach, J. C. (1997). Fynbos (fine bush) vegetation and the supply of water: A comparison of multi-criteria decision analysis and cost-benefit analysis. *Ecological Economics, 22*(2), 123–140.

Kemp, R., & Pearson, P. (2007). Final report MEI project about measuring eco-innovation, European Commision FP6, n°044513.

Komninos, N., Pallot, M., & Schaffers, H. (2002). Smart cities and the future internet in Europe. *Journal of the Knowledge Economy, 4*(2), 119–134.

Le Boennec, R. (2013). *Les mobilités urbaines: quelles interactions entre déplacements durables et ville compacte?* Nantes: Doctoral dissertation.

Le Boennec, R. (2014). Externalité de pollution versus économies d'agglomération: le péage urbain, un instrument environnemental adapté? *Revue d'Économie Régionale & Urbaine, 1*, 3–31.

Le Boennec, R., Nicolaï, I., & Da Costa, P. (2017, January 24–25). *Inciter au changement de comportement dans les pratiques régulières de mobilité: une analyse multi-acteurs multicritères*. Conference Paper, ATEC-ITS France.

Le Boennec, R., & Sari, F. (2015). Nouvelles centralités, choix modal et politiques de déplacements: le cas nantais. *Les Cahiers Scientifiques du Transport, 67*, 55–86.

Lera-López, F., Faulin, J., & Sánchez, M. (2012). Determinants of the willingness-to-pay for reducing the environmental impacts of road transportation. *Transportation Research Part D: Transport and Environment, 17*(3), 215–220.

Leydesdorff, L., & Deakin, M. (2011). The triple-helix model of smart cities: A neo-evolutionary perspective. *Journal of Urban Technology, 18*(2), 53–63.

Macharis, C., & Bernardini, A. (2015). Reviewing the use of multi-criteria decision analysis for the evaluation of transport projects: Time for a multi-actor approach. *Transport Policy, 37*, 177–186.

Macharis, C., De Witte, A., & Ampe, J. (2009). The multi-actor, multi-criteria analysis methodology (MAMCA) for the evaluation of transport projects: Theory and practice. *Journal of Advanced Transportation, 43*(2), 183–202.

Mackay, M., & Metcalfe, M. (2002). Multiple methods forecasts for discontinuous innovations. *Technological Forecasting and Social Change, 69*, 221–232.

Mahieu, P. A., Crastes, R., Kriström, B., & Riera, P. (2015). Non-market valuation in France. An overview of the research activity: Introduction. *Revue d'économie politique, 125*(2), 171–196.

Markides, C. (2006). Disruptive Innovation: In need of a better theory. *Journal of Product Innovation Management, 23*, 19–25.

Marks, M. (2016, October). *People near transit: Improving accessibility and rapid transit coverage in large cities*. Report for the Institute for Transportation and Development Policy (ITDP).

Marsal-Llacuna, M. L., Colomer-Llinas, J., & Melendez-Frigola, J. (2014). Lessons in urban monitoring taken from sustainable and livable cities to better address the smart cities initiative. *Technological Forecasting and Social Change, 90*, 611–622.

Mayor, K., Lyons, S., Duffy, D., & Richard, S. J. (2012). A hedonic analysis of the value of rail transport in the Greater Dublin Area. *Journal of Transport Economics and Policy, 46*(2), 239–261.

Meunier, V., & Marsden, É. (2009). *Analyse coût-bénéfices: guide méthodologique*. FonCSI.

Nemet, G. (2009). Demand-pull, technology-push, and government-led incentives for non-incremental technical change. *Research Policy, 38*(5), 700–709.

Pillot, J. (2011). *Vers une mobilité décarbonée: quels ecosystèmes d'affaire, quels positionnements stratégiques?* Rapport Institut Transition Energétique VEDECOM.

Pouyanne, G. (2005). L'interaction entre usage du sol et comportements de mobilité. Méthodologie et application a l'aire urbaine de Bordeaux. *Revue d'Économie Régionale & Urbaine, 5*, 723.

Quinet, E. (2010). *La pratique de l'analyse coût-bénéfice dans les transports: le cas de la France*. OECD/ ITF Joint Transport Research Center Discussion Paper, n° 2010–2017.

Rennings, K. (2000). Redefining innovation – Eco-innovation research and the contribution from ecological economics. *Ecological Economics, 32*, 319–332.

Repko, J. (2012). *Smart cities literature review and analysis smart cities* (pp. 1–18). Washington, DC: IMT 598 Emerging Trends in Information Technologies, University of Washington.

Rifkin, J. (2011). *The third industrial revolution: How lateral power is transforming energy, the economy, and the world*. New York: Palgrave Macmillan.

Rosen, S. (1974). Hedonic prices and implicit markets: Product differentiation in pure competition. *Journal of Political Economy, 82*(1), 34–55.

Rousval, B., & Bouyssou, D. (2009). De l'aide multicritère à la décision à l'aide multicritère à l'évaluation.

Tudela, A., Akiki, N., & Cisternas, R. (2006). Comparing the output of cost benefit and multi-criteria analysis: An application to urban transport investments. *Transportation Research, Part A: Policy and Practice, 40*(5), 414–423.

Twigg, J. (2009). *Characteristics of a disaster resilient community*. London: Department for International Development.

Utterback, J. M. (1994). *Mastering the dynamics of innovation*. Boston: Harvard University Business School Press.

Utterback, J. M., & Abernathy, W. J. (1975). A dynamic model of product and process innovation. *Omega, 3*(6), 639–656.

Verhoef, E. (1996). *The economics of regulating road transport*. Cheltenham: Edward Elgar Publishing.

Worku, G. B. (2013). Demand for improved public transport services in the UAE: A contingent valuation study in Dubai. *International Journal of Business and Management, 8*(10), 108.

Combining Public and Private Strategies Towards Sustainable and Responsible Mobility

Danielle Attias and Sylvie Mira Bonnardel

Abstract Urban capitals worldwide are experiencing huge mobility changes which engender deep modifications of the urban space. The entrance of newcomers and new mobility services are all based on the collective awareness that a sustainable economy cannot develop without smooth, ecological and sustainable mobility. Developing a sustainable city is becoming the main stake in the forthcoming massive urbanization. This transformation occurs by connecting private and public actors, as private actors can alone design neither innovative business models nor appropriate strategy without public actors. Partnerships between private and public actors are necessary, but also complex. Combining public and private offers for the new mobility is creating opportunities, but also constraints. The revolution in urban mobility aims to be intelligent and user-centered. This concept of mobility-as-a-service is based on an offer of mixed mobility is going to afford the city-dweller safer and more sustainable mobility. To study successful private/public partnerships builds the framework for a repositioning of the traditional actors and to a redefining of their role.

Keywords Sustainable mobility · Public/private partnerships · Corporate strategy in urban mobility

1 Introduction

Big cities and urban areas have been tremendously enlarging their frontiers these last 3 decades and clearly intend to continue their expansion during the forthcoming decades. A U study indicates that, prior to 2030, the population living in cities should

D. Attias (✉)
Laboratoire Genie Industriel, CentraleSupélec, Université Paris-Saclay, Gif-sur-Yvette, France
e-mail: danielle.attias@centralesupelec.fr

S. M. Bonnardel
Ecole Centrale de Lyon, Ecully, France
e-mail: sylvie.mira-bonnardel@ec-lyon.fr

© Springer International Publishing AG, part of Springer Nature 2018
P. da Costa, D. Attias (eds.), *Towards a Sustainable Economy*,
Sustainability and Innovation, https://doi.org/10.1007/978-3-319-79060-2_8

account for more than 60% of the world's population, with a forecast peak to 70% for 2050. This galloping urbanization urgently demands new urban policies and new lifestyles that could enhance the development of decarbonized transportation, carpooling, car sharing or ride sharing. Yet, current policies have yet to succeed in managing the increase in traffic and do not foresee opportunities and problems at stake.

According to Navigator Research,[1] smart mobility represents a huge market, covering infrastructures as well as services, which should increase worldwide from of \$5.1 billion in 2015 to \$25.1 billion in 2024. Smart mobility is now considered as the major preoccupation of urban policies linking economic development with population's wellbeing. Thus, urban transportation policies are simultaneously targeting decongestion and time reduction in moving safely, comfortably and less stressfully.

The question of new mobility that is smart, sustainable and responsible cannot be tackled only by public policies; it needs a plurality of strategies involving public and private actors. In this chapter, we aim at giving an overview on how partnerships between public and private actors are developing to design new answers to help implement a smart and sustainable urban mobility. In the first section, we are going to explain the complex mechanisms of public/private partnerships within the new mobility paradigm. Successful examples of partnerships in the world are detailed that show how the traditional roles of the private/public actors are transforming.

In fact, partnerships between public and private actors are no longer confided to the traditional task distribution between public authorities, on the one hand, investing in infrastructures, and private enterprises—and on the other hand, enriching it with intelligence and service. Indeed, within the new mobility, i.e. mobility as a service, public actors are no longer reduced to supporting the investment costs while private actors are not anymore limited to offer the services as two examples, Navya and Cyclopolitain, in the French city Lyon, demonstrate. We'll explain these two examples is the second section.

2   A Global Approach of Public/Private Partnerships

2.1   The Public/Private Partnerships Framework

In its widest definition, the term 'Public Private Partnership' covers all forms of association between the public sector and the private sector intended to implement all or part of a public service.

The terminology of Public Private Partnership corresponds historically to a new type of public contract created in England (Private Finance Initiative) in the 1990s and transposed into numerous countries. The public-private partnership has been

---

[1]Navigant Reserach, Urban mobility in smart cities, Juin 2015.

largely studied within the "new public management" (Marty et al. 2006) according to whom the budgetary difficulties which strike the public finances, burden the legitimacy of management of public services and require a renewal of the model. Two types of analyses were conducted: firstly the question of how to define the scope of intervention and, secondly, the techniques of resource mobilization, with the assumption that the private sector possesses and means which can aid governments to reach a higher satisfaction of the public services user.

The important development of public/private partnerships within urban mobility is due to the convergence of interests between, on one hand, the pressuring demand of the city-dwellers to improve mobility and the necessity of mastering the public finances and, on the other hand, the companies (could we replace 'the companies' with 'corporate'?) awareness of the forthcoming profitability of their investment.

The European Union has been largely supportive of Public Private Partnership to encourage the development of new mobility infrastructures, which cannot be carried only by public investments both in terms of reactivity and means (Van Miert 2003).

Many countries in the OECD have undertaken deep reforms of the way their public politics are being driven within a difficult budgetary context. This movement has been studied by new analysis such as the School of the Public Choice (Mueller 2004), the New Public Economy (Laffont 2000) and the New Public Management (Hood 1995)—all studies that have reinforced the new approach of the public policies (Perret 2001). In the macroeconomic vision, these reforms aim at transforming the global functioning of public policies by implementing new budgetary rules. At the same time, in a microeconomic perspective, these reforms are leading to the implementation of management and control tools converging with performance evaluation methods used in the private sector (English and Skellern 2005).

Public-private partnerships are facilitating the transfer of methodologies and accounting models from the private sector to the public sector. Besides they open the path for accountability, i.e. greater transparency and responsibility in the use of the resources. This convergence facilitates particularly the pre-evaluation of forthcoming actions: "The Public Private Partnership is well and truly a tool which allows to spend better, to improve the efficiency of the public policies", said Michel Destot,[2] President of the Association of the mayors of big cities of France (AMGVF).

Within the smart, sustainable, responsible mobility, public and private actors are forming partnerships to offer a new mobility—mobility as a service.

---

[2]In Courrier des maires de France Semaine 01/02/2013.

2.2 Mobility as a Service: Smart Combination of Public and Private Actions

Introduced in Sweden, the concept of mobility as a service (MaaS) was popularized by Sampo Hietanen, president of ITS Finland (Intelligent Transport Systems Finland), a non-profit organization which promotes a safer, more responsible, more sustainable and smart urban mobility. ITS Finland is a very good example of Public Private Partnership.

Widely covered in the media during Intelligent Transport Service Congress in Bordeaux in 2015, the concept of MaaS gave birth to the alliance of MaaS, accounting for more than 30 partners, public and private actors, among whom about 20 companies, aimed at developing fast, and on worldwide scale, smart urban mobility. Four working groups tackle questions concerning the development of new services, user needs, rules and technologies.

Concretely, mobility companies buy transportation services for transportation companies—who operate trains, buses, taxis, but also bikes or car hiring. The transportation companies are themselves customers of infrastructure and data.

Then, mobility companies sell MaaS through a smartphone application. The user indicates in the application his destination and his preferences; according to which the application immediately suggests a full itinerary composed of the different transportation services he could use (Fig. 1).

The moving process is thus largely simplified: only one application needs to be used for any urban mobility combining private services (such as taxi, Uber, car rental, car sharing, ride sharing, bikes hiring) and public services (such as train, subway, bus). The application will charge the user directly for the global service. This global combined approach is certainly far less expensive for the user than possessing a private car. Consequently, it is a major step forward for decarbonisation and decongestion in cities.

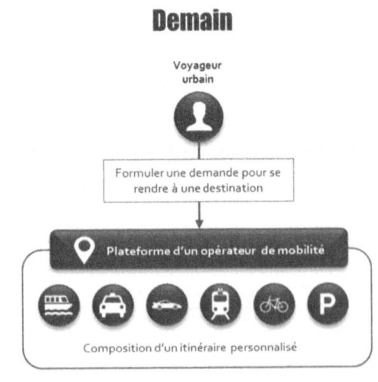

Fig. 1 Comparison offers of mobility today and in the future. Source: Ramphul (2014)

Key:

- *Today: The Urban Traveller;*
- *Buys a ferry ticket, takes a taxi, rents a car, buys a train ticket, rents a bike and pays for parking;*
- *Tomorrow: The Urban Traveller;*
- *Makes a destination request, which is received by a mobility operator platform, who then puts together a personalised itinerary.*

This transition towards a multi-modal mobility is not only due to sustainability objectives. It is also reinforced by inclusion strategies, i.e. the aim of improving the mix of lives and uses in the urban space (Clochard et al. 2008).

Urban policies must indeed be aware of the gentrification process and the need to ensure that gentrification of some the city's most vibrant neighborhoods does not displace the poor and elderly who want to stay in their communities, leading to the construction of urban *"citadels"* reserved for a new urban bourgeoisie (Guilluy 2014).

Moreover, there is also a movement to reconquer the urban space in terms of the automobile which has been overwhelming the space causing most of the city's problems (pollution, noise, safety, limitation of the vegetated spaces). This movement also takes place in the context of social mixing (Harvey 2011), and, in this context, multi-modality plays a major role.

As the cooperation between public and private actors has proved positive, the partnership initiatives has multiplied—a large number of which at the instigation of governments as indicates Center FUTURIS (2012) which details the existence of numerous cooperative structures in France. Among these are many clusters—Carnot institutes, thematic networks of advanced research and institutes of technological transfer. All these structures aim at boosting innovation, leading to entrepreneurship and job creation.

As Guicheteau and Millette (2012) demonstrated, the robustness and the balance of the partnership between public actors and private companies is the determining factor of the success of any project of sustainable mobility.

2.3 Examples of Successful Public/Private Partnerships in France and in the World

The new mobility presents a real melting pot of various successful partnerships between governments and private companies. These are either initiated by the government or by private companies, who aim either at facilitating vehicle flows through smartphone applications or allowing new mobility services.

2.3.1 The Case of France

The community of Saint-Quentin-en-Yvelines in the Paris area, together with the French car maker Renault, have launched a car sharing service called Twizy Way, for urban movements. Thanks to a specific smartphone application, the user can target the nearest vehicle and rent it immediately by flashing the QR code printed on the car.

In Lyon, the mobility platform called Optimod'Lyon is based on a new type of public/private partnership, private companies having invested in infrastructure while public research centers have dealt with data management. This partnership was conducted for a larger service range for the city's inhabitants, smoothening the flow of traffic in the city. Optimod'Lyon scans the city's traffic in real time, processing data from different private and public organizations, the regional government, the national railway company, local public and private bus companies, the subway company, the private highway companies.

The partnership has been gathering the city council with large private companies such as IBM, Orange, Renault Trucks, combining all competences and technologies to offer city dwellers an innovative platform to organize their urban movements.

The platform forecasts hourly traffic allowing for the optimization of 1500 crossroads with traffic lights of the urban area with the goal of improving the fluidity of road traffic. The browser on cell phones integrates a multimodal approach giving all possible scenarios of movement in the city with real time alerts.

2.3.2 The Case of Singapore

Singapore is also a good example of successful public/private partnership aimed at bringing visible improvements for its inhabitants' lives. The city developed a public program called *Smart Nation* based on the integration of digital technologies to solve urban problems. A governmental agency has been created that is fully dedicated to this question.

"We wanted to go a bit beyond the idea of the city as a machine—of 'smart' being applied to the city alone," says Jacqueline Poh, managing director of the Infocomm Development Authority of Singapore, which is leading the initiative. The Smart Nation is user oriented as says Jacqueline Poh: "The idea was to bring together different agencies to see: If we wanted to be a citizen-centric, business-centric, smart city, what really are the applications that would make sense? What would best define and improve the lives of our citizens and businesses? And then, working backwards, what are the kinds of technologies, what are the kinds of data that need to be collected and shared, and then made into tools to be able to enable that experience?

So to us Smart Nation is about the experiences that people live out in their everyday lives."[3]

According to Jacqueline Poh, this user-centered approach of the city has to be supported by the city's authorities in close collaboration with tech companies in order to rapidly design and implement solutions that correspond to reality.

2.3.3 The Case of the USA

The US Rust Belt represents another example of successful collaboration between local government and private companies. In their book 'The *Smartest Places on Earth*', Von Agtmael and Bakker (2016) explain how old industrial US cities (Detroit, Pittsburgh, Akron. . .) are becoming new innovation hot spaces, attracting young talented entrepreneurs resulting in a growing start-up settlement. One of the key success factors of the newly-reborn cities, the authors suggest, is all forms of collaboration between local startups, the remaining large companies, universities and research centers and the city governance. All these actors build a dynamic and innovation-oriented cluster. The clusterization of the local environment is helping the emergence of smart factories, with robots, cobots and drones working with workers. Companies and the city's government work together for a more attractive urban space by optimizing the transportation system and implementing new mobility services. The close collaboration between public authorities and private companies allowed for the Rust Belt to rise from the ashes. All of the above examples show that public/private partnerships can leverage the revolution of mobility, aiming an innovative, responsible and sustainable mobility. However, these partnerships do usually not emerge by themselves. They are either pulled by public authorities or sometimes pushed by the strategies of development of the private companies. In both cases, these partnerships are confronted with differences of culture, processes and organization, which imply difficulties execution, between regulation and negotiation.

The Sect. 3 of this chapter presents through two case studies the relations between private actors and public actors in the implementation of new forms of mobility.

3 Two Experiments of New Strategies for Urban Mobility

The public private partnerships are confronted with very diverse realities in the large cities of the world suffering from "communication distortions" (Habermas 1987), between public authority speeches and practical behaviors in the field of mobility. Public communications have been increasing the need to think about environmental

---

[3]Sidewalk Labs, http://www.atelier.net/trends/articles/smart-city-pouvoirs-publics-entreprises-collaborent-citoyen-sort-gagnant_442386

sustainability and people's health; but de facto, the users' behavior has evolved very little these last few years on the mobility question. How do we interpret this gap between a theoretically expressed will to experience urban movements differently and an ever-changing lifestyle?

Does this then formalize a *paradoxical injunction* (Watzlawick 1993) hindering the implementation of a sustainable and responsible mobility in large cities?

This question is central to our book. In fact, user-citizens live in a compulsory and not chosen frame of mobility because urban mobility is still limited by all kinds of boundaries, including local regulations. Speed limits, traffic reduction, parking spaces, risk of penalties—even bans, all design a new urban lifestyle where the pleasure of driving does not exist anymore.

The innovative experiment of mobility realized in the city of Lyon in France with a company called *Les Cyclopolitains* shows, on the contrary, users adopting a new mobility and enjoying a new way of surfing their city with pleasure. Nevertheless, the situation is still paradoxical, since some of their competitors do not hesitate to by-pass rules or regulatory measures, as we shall show it point 2.1.

When the public actor chooses another strategy, and agrees to make a commitment with private companies to invest new mobility forms such as driverless shuttles, the lifestyle in town radically transforms. Convergence between private interests and public interests are limited in Europe. But, when experiments of innovative mobility forms exist concretely on the field, all the public-private actors can enjoy substantial advantages. The convergence creates a positive, dynamic image of intelligent mobility and allows for a world of opportunities. This also brings financial outcomes for private companies, participating in the new mobility as we'll show in point 2.2.

3.1 Cyclopolitains, How to Bring New Mobility in a Competitive World

The company Cyclopolitains has been running its activity in Lyon, France, for 10 years. It supplies a new ecological and easy-to-use mobility service.

This new offer is seductive because the technology is simply based on a comfortable tricycle; the passengers and the driver are protected from bad weather by a hull wrapping the structure of the bike. The passengers sit behind the vehicle. The engine is electrically assisted—by a 250 watt engine—and subjected to safety standards and a speed limit (25 kph). The driver is present and has to maneuver the vehicle which can ride pedestrian streets, cycle paths and access the banks of rivers.

Vehicles move within the entire city and allow for tourists to have privileged access to monuments. But, they also fulfil a social function because they are relays for children, the elderly and shopkeepers situated in city center. The range of services offered by Cyclopolitains is attractive, for example, to fetch children at school, to deliver parcels, to accompany a customer to a meeting place or facilitate in-home services for old people. Over time, to transport people and the goods stood out as a profitable economic model. The company today counts 15 salaries and has recently developed cargo-bikes for express delivery.

How did Cyclopolitains obtain the authorization to circulate in the city center? How did this self-financed start-up, without any particular support, succeed in the urban landscape? The regulations for new mobility services are complex and it took a while before receiving all authorizations. But *the present decade is rather favorable to new offers; there is the beginning of consciousness for the need for mobility change. Pedestrian space has been taking over car space. By chance, the City government of Lyon has really supported the development of our project* says Sarah Dufour, Founder and President of the company.

Nevertheless the founders of Cyclopolitains are conscious of difficulties due to the regulations and the limits of political speech. A legal opportunity has helped them. As Dufour further explains, *our tricycle is considered as a bicycle and this label allows the access to pedestrian ways and other paths that are reserved for bikes and for rivers banks. The absence of regulations was, in the end, good fortune for us. In that case we can say that innovation laid down the law*.

However, in Paris, the situation for Cyclopolitains is totally different. Indeed political speeches on ecological vehicles were spreading in the French capital while the World Conference on the Climate (COP21) was happening. But Cyclopolitains has been confronted with unfair competition. Similar vehicles circulate with four times more powerful motorizations, prohibitive price rates and imported tricycles which allow these new competitors to realize consequent profit margins and disturb the market.

The complexity of local regulations is enhanced by the fact that the City government allowed the local authority and the police to regulate the traffic of tricycles. As we know it, the police have many more concerns than to regulate these vehicles and consequently there is no regulation in the French capital.

The situation in other cities of the world is different. In Berlin, Barcelona, New York, regulations are clear and allow for a profitable economic model. The companies whose tricycles do not conform to standards are strongly sanctioned.

3.2 Navya: An Audacious Experiment with Autonomous Shuttles in the Urban Space

Autonomous vehicles are a major new technological rupture. Very soon, cities will host the first driverless cars. Studies even forecast that by 2055, autonomous vehicles will exceed conventional vehicles in number. This market, within the estimates of billions of dollars in the years to come, is therefore a strategic issue of the twenty-first century. It embodies the revolution, set the technological foundation for the world of tomorrow and opens new perspectives.

In medium sized cities, autonomous vehicle experiments already exist. The French company Navya, created in 2014, develops driverless electric collective vehicles. This society, which is defined by its founder as a company that is specialized in the development of innovative, intelligent and sustainable mobility solutions, is building vehicles for the intelligent transport of people and goods. These vehicles are equipped with embedded technologies and multi-captors to interact with their environment. Also intended to integrate efficient mobility services, in terms of space and energy, these vehicles are built to allow sustainable inter-modality and multimodality. The motto of Navya is: "*a step ahead of the autonomous and electric vehicle technology*".

But the central question of these vehicles is to find a testing space to give visibility and credibility to the project. How did Navya circulate its shuttles at the heart of the city of Lyon? What were the margins of negotiation between the public and private partners?

Christophe Sapet, President of Navya, remembered that his fist concern was: "*to convince and reassure*". He has to prove that the urban collective transport system of tomorrow will be a real lever in improving the quality of life of urban residents, fulfilling their growing needs in travel with accessibility of all areas, support for people with disabilities, 24-h availability, reduction of noise and pollution.

It is in the area of Confluence (in the South of Lyon, France) that the first shuttles were tested in September 2016. The circuit is not closed since it connects leisure and shopping centers and the peninsula of Lyon by riding a dock along the Saone River. The shuttles are autonomous and their engines—which are 100% electric, they are free of charge and travel at 25 mph on a distance of 1.3 km. This large mainly pedestrian area offers a few services accessible to all mobility and the Navya shuttle has the advantage to answer a need for mobility for all types of users (employees, family and elderly).

However, this experience has been implemented after many negotiations between the Union of public transport Lyon (SITRAL) who saw in Navya, a direct competitor to its bus lines, and the city of Lyon, represented by the Mayor who was convinced and prepared to support this initiative. It should be noted that Navya had also obtained the support of Keolis, the first private operator of transit in France and ADEME (environment and energy control agency).

The arguments put forward by Navya to present its vehicles are numerous: "*without a driver with an electric engine, clean and responsible, the bus of the future will be more frequent and less polluting, less costly and more efficient. Indeed, free transport today is an extraordinary publicity for the shuttles; their regular frequency—every 10 minutes—is also a comfort for users. It is an efficient means of mobility. Finally, it is a green, clean and quiet vehicle*".

However, developing new technologies at the service of safety and security is a fundamental issue, particularly in the sector of transport of goods and passengers. Navya needs firstly to ensure optimal transportation safety: "*any shuttle accident would have a catastrophic impact on our image and our business. However, autonomous vehicles are sure to reduce accidents and increase road safety as we know that human error is responsible for 93% of accidents*", says Hervé Gentil, commercial Director of Navya.

This public-private partnership was a particular success in this case since Navya has benefited from its experiment in Lyon raising funds of 30 million € in 2016 and could also develop new experiments worldwide (in Switzerland, in the USA, in Singapore, in Australia).

For the Lyon city government, this is a positive local image and a window on the world. It is also an exemplary framework, of virtuous economy which, will be imitated in other cities.

At a time where the projects for driverless technologies are multiplying in the world, where new services of transportation of people and goods are developing, the relevance of autonomous and electric vehicles has not to be proven anymore. Do autonomous shuttles participate in major cities to the sustainable transformation of ways of life by reinventing mobility? This is the wager of the City of Lyon.

4 Conclusion

Whereas the car is still the main mode of transport for residents, city policies are moving towards reducing pollution and fostering the decarbonization of town centers. This vision of a healthier city is present in many projects including eco-neighborhoods, for example, where the car disappears from the urban space—thus permanently changing patterns of mobility is assumed to change the life of individuals radically. A real Metropolitan revolution is occurring and participating in the development of smart cities around the world.

In parallel with these changes, we are witnessing a real boom of innovation in many areas and sectors: digital transition and digital revolution, exploration and exploitation of new clean and renewable energies. Advanced technologies developed to design the vehicles of the future (GPS location, visual recognition, remote sensing) allow to develop other areas, such as robotics. Electric or autonomous vehicle research gives thus rise to new innovations, which meet the challenges of a connected and constantly evolving world.

References

Clochard, F., Rocci, A., & Vincent, S. (2008). *Automobilités et altermobilités: quels changements ?* Paris: l'Harmattan.
English, L., & Skellern, M. (2005). Editorial of the special issue on public-private partnerships. *International Journal of Public Policy, 1*(1–2), 1–21.
Guicheteau, J., & Millette, L. (2012). *Projets efficaces pour une mobilité durable: facteurs de succès*. Paris: Presses inter Polytechnique.
Guilluy, C. (2014). *La France périphérique*. Paris: Flammarion.

Habermas, J. (1987). Théorie de l'Agir Communicationnel. In *L'espace politique* (Vol. Tome 1). Paris: Fayard.

Harvey, D. (2011). Le capitalisme contre le droit à la ville – Néolibéralisme, urbanisation, résistances. Éditions Amsterdam.

Hood, C. (1995). The new public management in the 1980s: Variations on a theme. *Accounting, Organizations and Society, 20*(3), 93–109.

Laffont, J. J. (2000). *Incentives and political economy*. Oxford: Oxford University Press.

Marty, F., Trosa, S., & Voisin, A. (2006). *Les partenariats public-privé*. Paris: La Découverte, Repères.

Mueller, D. C. (2004). Public choice. In C. K. Rowley & F. Schneider (Eds.), *The encyclopedia of public choice* (Vol. 1, pp. 32–48). Dordrecht: Kluwer Academic Publishers.

Perret, B. (2001). *L'évaluation des politiques publiques*. Paris: Édition de la Découverte.

Ramphul, Y. (2014). Smart city: la mobilité as a service en test à Helsinki. Habitants, News, 20 Août 2014.

Van Miert, M. K. (2003). Groupe à haut niveau sur le Réseau Transeuropéen de Transport. Rapport à la Commission Européenne.

Von Agtmael, A., & Bakker, F. (2016). *The smartest places on earth: Why rustbelts are the emerging hot spots of global innovation*. New York: Public Affairs, Perseus Book.

Watzlawick, P. (1993). *The language of change, elements of therapeutic communication*. London: W.W. Norton and Company.

Opening in Conclusion: An Anthropological Approach to Transformation—Shaping Shapes

Angela Minzoni

Abstract Since D'Arcy Thompson's pioneering work On Growth and Form published in 1942, continuous efforts have been made in Western academic fields to understand and capture forms and evaluate their similarities and differences, their continuities (invariants) and their discontinuities through time. Mathematics has been used intensively to capture forms such as those of cells, gases and animal or plant shapes. It has also been used by anthropologists to capture what has been interpreted as the forms or structure of languages, myths and symbols. That is, mathematics has supported the attempt to give a form to non-tangible, immaterial human expressions and thoughts. Organizations and companies also feel the need to define their structure or the shape taken by their relationships and interconnections. Here again, mathematics is at work in this attempt to give a form to how they function. This paper focuses mainly on the Western quest for form and structure and gives some examples of the methods used to address this quest. In so doing, it formulates the following interdisciplinary research question: can we think about the transformation of functions which do not have any intrinsic shape? How does the idea of "environmental shape" captures the function of sharing in order to renew our perception of sustainable development?

Keywords Living systems · Biology · Evolution · Fuzzy models · Functional approach · Environment

1 Introduction

Transformation is first of all an action, that of 'trans-forming'. But it is a continuous action, performed both by us and on us. Yet not all changes are necessarily a transformation, because transformation implies a change of form. Anatomy,

A. Minzoni (✉)
Laboratoire Genie Industriel, CentraleSupélec, Université Paris-Saclay, Gif-sur-Yvette, France
e-mail: angela.minzoni@centralesupelec.fr

© Springer International Publishing AG, part of Springer Nature 2018   135
P. da Costa, D. Attias (eds.), *Towards a Sustainable Economy*,
Sustainability and Innovation, https://doi.org/10.1007/978-3-319-79060-2_9

physiology, mathematics and physics have drawn on the concept of transformation since the Middle Ages. However, it was not employed by organizations until much more recently, mainly since the nineties when firms started facing challenges such as mass customization, supply chain management, corporate downsizing or virtual organization all related to the omnipresent and significant growth of information systems.

In the West, the persistent questioning within various disciplines regarding form, and transformation, is striking. Mathematics has examined the shape of the circle and the triangle. Physics has studied the shape of trajectories of bodies in motion. Chemistry stems from questions about the shape of solids and their metamorphosis into liquids and gases. And in biology, the shapes of living beings are the focus of investigation.

Transforming is not to be confused with 'reforming', the primary meaning of which is to return to a primitive form or shape, rather than give rise to a new shape, which is the case with transformation. Nor with 'improving' because, in this case, the changes aim simply to maximize the performance of what already exists, according to certain constraints, often related to cost.

The fundamental question raised by the concept of transformation is how we determine at what point we can say that a shape has changed. That is, that the change in morphology involves a change in function. How can we characterize shapes in order to evaluate whether they have changed? Is transformation a break or an evolution? How is the action of transforming triggered? What is the role of the environment in the transformation of organisms? And what is the role of organisms in the transformation of the environment?

2 Pioneering Thoughts About Shapes

It is to Thompson (1917) that we owe some of the most notable interdisciplinary thinking on the subject of transformation, as well as the first 'theory of transformations'. A biologist, mathematician and historian, Thompson wrote a fundamentally important book: On Growth and Form (1942). In it, he analyses living beings from the perspective of a geometrician and helps us understand through numerous illustrations the elements of transformation: the dimensions and scales, volumes, speed and temperature. He thus relies primarily on mathematics to define shapes and their deformations. He also paid great attention to invariants, typically the details which are maintained from one stage to another of a given shape, despite the transformations. To illustrate this, he drew a squared grid (see Fig. 1) to represent the system of local or particular coordinates. The grid then becomes curvilinear to represent the subsequent stages of transformation, while retaining the same coordinates. This method helps us understand the continuities and discontinuities in the process of transformation from one state to another for a given system or living organism.

Thompson's precursory questioning paved the way for methods for identifying and understanding shapes, how they function, their outlines and how their

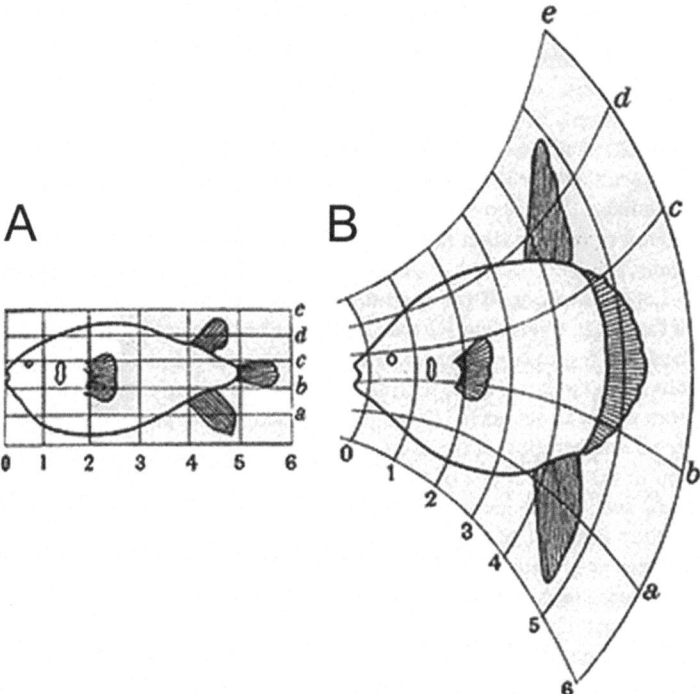

Fig. 1 A diodon, which transforms into a sunfish (On growth and form: p. 749)

component parts harmonize and coexist. He believed that there is an underlying logic to transformations, which cannot be violated. He thus endeavoured to reduce the variety of shapes to a general schematic and propose methods to quantitatively measure differentiation.

In 1950, almost 30 years later, Alan Turing (1952) picked up on the question raised by Thompson and proposed a general theory of morphogenesis. He attempted to mathematically model the chemical processes at work in the formation of patterns in freshwater hydra, which have the ability to spontaneously divide into two or three parts, from which complete new creatures can regenerate. Turing surmised that this regeneration is the result of interactions between two chemical compounds, one playing the role of activator and the other inhibitor, which he reduced to a system of linear differential equations. The solutions to these equations yield six potential cases where attachment plaques appear, from which regeneration takes place. His model, which indicates that genes control the speed of the reactions involved, helps us understand the rich variety of structures, which can be explained by a minimal set of purely physical-chemical mechanisms.

The work of Thompson and Turing was an attempt to find the structure underlying the diversity of shapes. Their respective work has had a huge influence on research approaches in other disciplines, which have also sought to determine these underlying structures, including in areas where shape is absent, such as a story or a belief, for example. Both are forms of expression based on the voice and the emotion conveyed by it. In addition, neither have limits or delimitation and consequently have no form or shape. To support this quest for underlying structure, mathematics has been used with increasing frequency.

In attempting to apply the concept of shape to things with no inherent morphology, the idea has emerged that shape is not a physical attribute of objects alone, but can also encompass the notions of process or structural organization. Here again, we find the concept of (underlying) structure, while the word 'shape' is used to refer to both the shape of a skull or a fish, and equally to the shape of an object, a machine or an idea. However, in the case of machines or objects, their shape depends on their function, determined externally, whereas in other cases, such as biological shapes, are characterized by self-organization and emergence processes. A machine, like each of its parts, is designed to perform a predetermined function. The shape of a living being is the result of morphogenetic interactions, which are internal to the organism itself, and is therefore innate. The morphogenetic processes of living beings are natural, autonomous and spontaneous in character. The form of a living being owes almost nothing to the action of outside forces (Monod 1970).

Since prehistoric times, the human imagination has given rise to mythological 'shapes', transforming what already exists by distorting the size or shape of animals or plants, for example, multiplying or removing an existing part, or even recombining parts. In an attempt to understand these expressions of the imagination, anthropologists developed a structuralist approach in order to identify the constraints to which these imaginary shapes are subject, whether found initially in the form of sketches, or later with oral or written description. The structuralist method enabled them to observe the appearance of convergences in the way different cultures repeatedly produce identical shapes. According to anthropologists, the function of these imaginary shapes and their transformations is to express and convey emotions as well as to develop the ability to communicate. Communication and emotion are adaptive properties of living beings. Living, surviving and communicating are all realms specific to emotion.

The structuralist approach in anthropology inextricably links the concept of transformation to the concept of system, transformation being the transition of a (social) system to another by modifying certain elements while retaining the same structure (invariant elements). Structural anthropology thus adopts the mathematical and biological concepts developed by scientists like Thompson or Turing and makes comparative approaches a key part of its method in order to detect, by comparison, the similarities (invariants) between different structures and, ultimately, their internal cohesion. According to Lévi-Strauss and Eribon (1990), an emblematic figure of this movement, to talk about structure, invariant connections need to form between the elements and the relationships of multiple entities, such that we can transition from one entity to the other by means of a transformation. In his studies of myths, and given the ability of mathematics to model relationships and evolutions, Lévi-Strauss even developed an equation intended to illustrate the symmetrical and inverse nature of transformations in the construction of myths:$Fx(a): Fy(b): Fx(b): Fa\text{-}1(y)$. (1958). The relationship $Fx(a): Fy(b)$ links elements belonging to a given domain, while $Fx(b): Fa\text{-}1(y)$ links two elements from different domains, $Fx(b)$ belonging, like the first two members of the formula, to a first domain and Fa-1(y) belonging to a second.

In the late 1950s, structuralism, in search of a general science, seemed to offer the promise of a reconstruction of knowledge, beyond the dichotomy of science/literature. Such a holistic approach to structure was most widespread in Eastern Europe and France, whereas in the Anglo-Saxon countries a much more descriptive and empirical method gained ground, such as the functionalism pioneered by Bronislaw Malinowski (1884–1942), for example. Malinowski's functionalism holds that, in a culture, each element has a function, comparable to that of an organ in a living body, and meets a need. It is these individual organic needs, which are the subject of transformation. As such, they transform into social imperatives.

Compared to the structuralist comparative method, Malinowski's functionalism developed a method that makes it possible to interconnect the various elements observed and, in turn, give them an overall cohesion. His fieldwork focused, for example, on the systems of exchange between tribes of the Melanesian Islands, particularly the more symbolic exchanges, used to establish political alliances. It was thus essentially the function of a culture that interested Malinowski. This is a different field from linguistic analysis, for example, a typical focus of study for the structuralists, and which can be understood via combinatorial analysis of a limited number of logical possibilities.

In line with Malinowsky's functionalism, which we might call 'absolute functionalism', Robert Merton (1949) developed a 'moderate' or 'relativized' functionalist theory and proposed two new categories: latent function and latent dysfunction. This author also stressed the need to develop 'medium-range theories', in other words limited to a particular problem, presenting concepts with a lower level of abstraction, in a break with the major overall theories of social functioning. He showed that systems have internal contradictions and that the study of a system must seek to understand what regulates, what maintains, what deregulates and what destroys relationships between individuals. He also demonstrated that the consequences of actions may be very different from those intended or expected.

The functional approach seeks to determine the movements that make it possible to understand and subsequently maintain the balance of a system, with a particular emphasis on action, usage and utilization. It is based on the concept of tension. That is, what hinders stability and retroactive effect, or the repercussion of an effect on its own cause.

The debate between structuralism and functionalism could be summarized as that of form and function. While the study of physical bodies can be readily understood by their shapes, the study of social 'bodies' lends itself less well to being understood in terms of shape.

The Roman Empire was already using the metaphor of 'social body' to convey the idea that society, like the physical body, cannot survive if its members are disconnected. This metaphor served as a call to cohesion. There is a clear determination to transpose to social bodies the same formal approach that puts structure at the centre of how we understand a system. The concept of trans-formation would thus be closer to shape and structure than to function.

3   Tracking the Evolution of Shapes

Although organizations in general and companies in particular take only a passing interest in anthropological debate, they are nonetheless subject to the impact of such debate, albeit indirectly. For example, those responsible for strategy raise the question of transformation and its limitations: how far is it possible to vary the scope of a system without it breaking or losing its shape? The answer to this question is often found by focusing on structure (largely driven today by Information Technology), with the idea that function (driven by humans) will duly follow and adapt.

The problem that remains unresolved, including for companies, is that of the evolution of these shapes and how it can be traced in order to evaluate whether what is taking place is a transformation, or another type of change, without the '-trans' prefix. In a changing world, it is important to understand exactly what kind of changes are taking place. Few of these changes could be called transformations. It is useful here to note the recent importance attached to ephemeral spaces such as pop-up stores, instant cities and light buildings (Ferreira da Souza 2014), which are closely tied to micro-local contexts and which, to a certain extent, give a new perspective to the concept of 'trans': that of ensuring a duration of occupation of the space, but in different ways. The political and economic ability to organize the ephemeral seems to have become the basis of sustainability, while at the same time ultimately altering our perception of continuity.

It is also important to identify what a trans-formation could mean when it applies not to form but to function. Can function take precedence over form? A positive answer to this question has been provided by architects such as Mies van der Rohe, Gropius and Le Corbusier, who have shown that the shape of buildings is dictated by their function, or the use we want to make of them. But what can be done if usages change faster than it is possible to change the structure of a building? Recent thought seems to point to the development of shapes that can better adapt to new functions through time, although always within given functional parameters (Brunet and Contré 2014). This is consistent with Merton's 'medium-range theory'.

Another important difference can be observed in the area of decisions to transform a shape or, if we can still talk about transformation for functions, to transform a function. Social 'bodies' can be seen as living organisms able to transform by themselves, from the inside, with no deliberate decision to transform taken externally. Attempting to provide a framework for understanding transformations of the immaterial, such as knowledge, motivation or learning, probably requires us to revisit the link between form and function, while not confusing the two or emphasizing one over the other. Maybe we will need to ask the question in terms of temporality, since in our current perceptions shape is more durable than function. However, at the same time, we are constantly surprised at the extraordinary longevity of functions, or some of them at least.

Among the various methods for approaching shapes and functions, models offer a way to understand them and, more importantly, to interconnect them, while always taking care to define a perimeter. The concept of 'pattern recognition' (Nagy 1968)

Fig. 2 Giving shapes to
service's features (a, b)

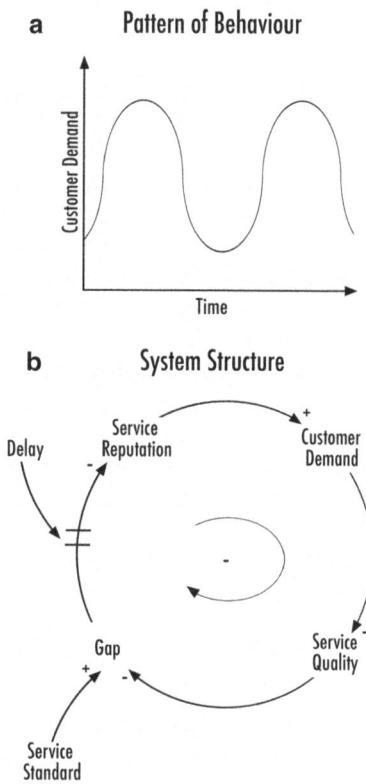

a   **Pattern of Behaviour**

Customer Demand

Time

b   **System Structure**

Delay   Service
Reputation

Customer
Demand

Gap

Service
Quality

Service
Standard

helps us understand the arrangement between the parts that form a whole, and those parts may come under the category of form or function. How parts belong to a whole and the links (typically links of influence) that unite them is emphasized in certain 'fuzzy' modelling methods. In addition, an element can belong to several parts, but in proportions and temporalities that are specific to each pattern. A pattern can be considered as a design where structures and functions are interlinked. The philosophy of 'fuzzy logic' ties in with research on 'situated knowledges', which emphasizes the importance of the specifics of each situation at a given time (Haraway 1988).

To reinforce the formal representation of social or organizational situations, mathematics is once again called on by C. Kirkwood (1998) and other authors as an attempt to 'capture' the shape of a service function (see Fig. 2a, b).

Fuzzy logic involves a non-binary approach, the uncertain nature of which reflects the uncertainty of the system itself. This logic seems particularly appropriate as a way to align with a functional approach, where usages and roles may evolve, invert and be modified from one user to another and over time. It allows us to bring together the theoretical aspects and the empirical aspects, as understood and experienced by the modellers of a given system. Picking up on the approach developed by R. Axelrod (1976) for the construction of cognitive maps, and adding to it the calculus methods derived from the study of formal neural networks, the basis of which was developed by McCulloch and Pitts (1943) as part of their work on cybernetics, driven by the Macy Conferences, new approaches have emerged since the end of the 1980s (Kosko 1986).

This type of model, with different levels and kinds of granulometry and input data, seems well suited to contexts that attempt to understand both form and function via the establishment of networks that are modified by the modification of their connections. Here, by analogy with biology, the connections are represented and a weighting is given to them, like the axons that link neurons together. This makes it possible to simulate the relationship between neurons, depending on whether this relationship is activated or not. These attempts are progressively moving toward the establishment of a concept that may, for functions, be the equivalent of the concept of transformation for shapes, while retaining the functional logic. The development of neurosciences, which owes much to the ability to produce pictures of brain function, may also help emphasize the benefit of representing the shape of functions, which is often overlooked today, when most studies focus on the function of shapes.

This will also be an opportunity to give the common points and similarities (invariants), which endure irrespective of trans-formations, the same attention that has been given in the last few decades to variances and differences.

4   Shaping Environmental Shapes: Starting a Reflexion About "Growth and Function "

The reflexion 'On Growth and Function' will open a new path for future interdisciplinary research aiming to capture the intangible representations at the base of spatial organisation or object production. Environmental issues can be considered as a promising starting point for such a topic. Especially when new approaches (Blanc 2016) begin to introduce the concept of "environmental shape" where politics and aesthetics are bind together highlighting the importance to be given to the lived experience. Here the environment is no more considered as an exogenous phenomena but is understood as a part inside the living, inscribed in the form of the living. Within this frame of thought the environment acquires a new status: the one of public common good, accessible to all where everyone can reclaim his real life space. There is a tight link between the sharing of resources, the form we give to our environment and the creativity or innovation.

The environment is a limited resource and sharing limits is probably our biggest challenge. Contrary to the late sixties (Hardin 1968), we can now consider new ways of sharing rather than the only State control or privatization of resources. Experimental research has pointed out the need to strengthen collective action theories starting from collective experiences. Among the criteria identified by Elinor Ostrom (1990) for sustainable management of common resources, we highlight here the first one: the need to make a clear definition of the contents of the common pool resource and effective exclusion of external un-entitled parties. This advocates for the resources to be given a shape: both a physical and an organizational shape but also a cognitive and an aesthetic one.

5 Conclusion

The collective elaboration of these shapes calls for the essential sorting out of the data to be put at work, for the innovative design of social, economic and political models and for the taking into account of the sensitive, embodied experience. In shaping the environmental resources we reshape our mental representations about them who in turn will shape the environment in a new way and under new circumstances. We are at the beginning of a new trans-formation cycle linking autonomy and awareness of ecological interdependencies together.

References

Axelrod, R. (1976). *The structure of decision: Cognitive maps of political elites*. Princeton: Princeton University Press.

Blanc, N. (2016). *Les formes de l'environnement*. Genève: Metis Presses.

Brunet, J., & Contré, O. (2014). Le concept du monospace: la simplexité dans la construction architecturale. In *Complexité-Simplexité, Actes du Colloque* (pp. 179–186). Paris: Collège de France.

Ferreira da Souza, R. C. (2014). *Ephemeral spaces*. Sheffield: University of Sheffield.

Haraway, D. (1988). Situated knowledges: The science question in feminism and the privilege of partial perspective. *Feminist Studies, 14*(3), 575–599.

Hardin, G. (1968). The tragedy of the commons. *Science, 162*, 1243–1248.

Kirkwood, C. (1998). *System dynamic methods*. Arizona: University of Arizona.

Kosko, B. (1986). Fuzzy cognitive maps. *International Journal of Man-Machine Studies, 24*(1), 65–75.

Lévi-Strauss, C., & Eribon, D. (1990). *De près et de loin*. Paris: Odile Jacob.

Lévi-Strauss, C. (1958). *Anthropologie structurale*. Paris: Plon.

McCulloch, W., & Pitts, W. (1943). A logical calculus of ideas immanent in nervous activity. *Bulletin of Mathematical Biophysics, 5*, 115–133.

Merton, R. (1949). *Social theory and social structure*. New York: The Free Press.

Monod, J. (1970). *Le hasard et la nécessité*. Paris: Seuil.

Nagy, G. (1968). State of the art in pattern recognition. In *Proceedings of the IEEE* (Vol. 56, pp. 836–862).

Ostrom, E. (1990). *Governing the commons. The evolution of institutions for collective action*. Cambridge: Cambridge University Press.

Thompson, D'A. W. (1917). *On growth and form*. Cambridge: Cambridge University Press.

Turing, A. (1952). The chemical basis of morphogenesis. *Philosophical Transactions of the Royal Society of London, Series B, Biological Sciences, 237*(641), 37–72

The manufacturer's authorised representative in the EU is Springer
Nature Customer Service Centre GmbH, Europaplatz 3, 69115 Heidelberg,
Germany. If you have any concerns regarding our products, please
contact ProductSafety@springernature.com

Printed and bound by CPI Group (UK) Ltd, Croydon, CR0 4YY
29/04/2026
02099459-0016